CALIFORNIA
GOLD

CALIFORNIA GOLD

THE BEGINNING OF MINING
IN THE FAR WEST

By

RODMAN W. PAUL

A
BISON
BOOK

UNIVERSITY OF NEBRASKA PRESS · LINCOLN

First Bison Book printing: March, 1965
Second Bison Book printing: April, 1967
Third Bison Book printing: March, 1969

PRINTED IN

THE UNITED STATES OF AMERICA

TO
Frederick Merk

When the original edition of *California Gold* went on sale in 1947, the West was deep in a series of centennials, one of which was intended to commemorate the great rush that started at Sutter's Mill in 1848. By the time the observance of that centennial was ended, those who had been active participants were as exhausted as if they had been digging gold themselves, while even those who had played only a quiet role were taking vows never to read another new book that mentioned Forty Niners, gold dust, or vigilantes.

The mood passed. Enthusiasm gradually returned, bringing with it the realization that along with the too frequent ceremonies and the too numerous publications had come permanent gains of a highly constructive sort. Scattered through the mass of material that the printing presses had spewed forth so undiscriminatingly were books that not only were valuable because of their own content, but that had the further virtue of suggesting new lines of inquiry that would challenge other writers for years to come. Correspondingly, out of the many local celebrations and "dramatic re-enactments" of historic events had emerged enduring projects, such as the state's successful restoration of the old mining town of Columbia, and its comparable development of the gold discovery site at Coloma into a historical park. Less permanently, old buildings throughout the gold rush country had been identified, studied, and marked, so that at least later comers would know what crimes they were committing when they bulldozed fine old structures into rubble in order to construct freeways and filling stations.

Today, if anyone desires proof that there is a widespread and enthusiastic interest in California's gold days, he need only drive along the appropriately numbered Highway 49, which traverses the Mother Lode country and Nevada, Yuba, and Sierra Counties. On any weekend from spring to autumn thousands of sightseers crowd the narrow, hilly streets of Nevada City, Grass Valley, Auburn, Placerville, Jackson, Sonora, and other former mining towns, while Colombia and Coloma are filled with visitors.

In these weekend crowds, tourists for whom this is an entirely new interest mingle with mining buffs who have given years of close attention to the history of the foothill gold towns. This edition of *California Gold* might well be dedicated to the mining buffs and the *aficionados* of California history, for it was partly in response to pleas from these avid students of California's past that arrangements were made to put the book back into print.

In the meantime our scholarly understanding of California's golden era has continued to mature, as professional historians and some highly competent amateurs have kept discovering additional manuscript records that give us new evidence concerning the lives of the gold seekers. Some of these new materials, such as the highly detailed autobiographical account of Herman Reinhart that Doyce Nunis edited, and the journal and letters of John Doble that

PREFACE TO BISON BOOK EDITION

Charles L. Camp prepared for publication, would have enriched *California Gold* if they had been available when the book was written twenty years ago. But they do not materially alter the conclusions reached at that time; rather, they confirm and document deductions that were arrived at on the basis of the less plentiful evidence then known.

The present edition is a reprinting of the original text without alterations. To it I have appended a brief and highly selective supplementary bibliographical essay. The essay does no more than suggest which of the new books seem especially important for the particular purposes of this volume. As an explanation of what those purposes are, the preface to the original edition still stands as a full statement.

RODMAN WILSON PAUL

California Institute of Technology
Pasadena
October 1964

PREFACE

A proposal to tell the story of mining for California gold suggests the aphorism concerning the weather: people are always talking about it, but no one ever does anything about it. Although much has been said about the romance and importance of the search for gold, few authors have sought to explain how the gold miner worked and lived, where he learned his craft, whether he profited from his labor, or how he managed to survive after the passing of the often described flush days of 1849.

This book is an attempt to answer those questions. It is a study of the twenty-five years of California gold mining that began with Marshall's discovery in 1848 and ended in 1873. Within that twenty-five-year span, California gold mining passed through successive stages of development that in other countries and in other industries would have required generations. The contrast between the exuberant youth of 1848-49 and the sober maturity of the early seventies could hardly be greater. A world of social, economic, and technical change lies between the two.

If one is to grasp the full significance of these changes, one must consider the background against which California mining developed. California was the first part of the United States to undertake precious metal mining on a large scale, and for that reason was the school in which the builders of the mining west learned their lessons. During the great boom of the sixties California-trained miners migrated to all parts of the Far West, carrying with them the knowledge they had acquired in the Golden State. But only a few of the methods used in California were peculiarly Californian in their origin. California mining, far from deriving its inspiration solely from the local scene, was the product of a complex blending of ideas, techniques, and human efforts contributed by many parts of both the Old World and the New.

It is important also to realize that California, in common with several other regions of the mining west, had an early development which deviated sharply from the American norm. Historians have long been familiar with the succession of pioneers who settled most of the early "wests." First came the fur trader, then the frontiersman who lived by hunting, fishing, and grazing his few head of livestock, then the pioneer farmer with his limited agriculture, then the farmer who cultivated the soil more intensively, and, finally, the builders of cities and towns.

In California this sequence was broken after 1848 by the intrusion of the new mining industry. The peripatetic prospector took the place of the frontiersman as the first tester of unsettled areas. The working miner supplanted the pioneer farmer as the first developer of the regions discovered by the prospector. The newly rich mining man and the San Francisco or London capitalist undertook the final task of intensive development. Meanwhile towns and cities grew up overnight, and died almost as quickly.

For the present writer this account of California mining forms a part of a more general study, now well under way, of the economic and social evolution of California during the twenty-five years that began at Coloma in 1848. Eventually perhaps the field of work can be expanded to include the whole Far West during these early years. For the moment it seems wiser to concentrate on the problems of a single state before venturing into a wider realm.

This volume has had to await publication for a longer time than the author ever intended. The research for it was done partly in the East, from 1938 to 1940, but primarily in California during 1940-41, while studying under a Sheldon Traveling Fellowship granted by Harvard University. The manuscript was ready for final revision and retyping late in 1942, when wartime interruptions intervened.

Before reaching the stage at which he felt justified in attempting a book, the writer served an apprenticeship under a notable trio of teachers of American history. Throughout three memorable undergraduate years it was his good fortune to have for a tutor Paul H. Buck, now Provost of Harvard University and Dean of the Faculty. As a graduate student the writer enrolled in the courses given by Arthur Meier Schlesinger, and began a fairly regular attendance at the Schlesingers' famous Sunday afternoon teas. During the years since then there have been many occasions for seeking guidance from Dean Buck and Professor Schlesinger, but both men have been as patient in giving advice as they have been wise in their counsel. Frederick Merk taught the writer during the latter's undergraduate and early graduate days. He then agreed to direct the author's later graduate studies and to supervise the preparation of this book in its original form. He has proved to be the kindest of supervisors and the most inspiring of teachers. It has been impossible to work under him without catching something of his enthusiasm for his subject and without coming deeply to appreciate his unselfish willingness to give of his time to his students.

When once started on the preparation of this book, the writer began to receive help from many Californians who were interested in the history of their state. The staffs of all the great California libraries were very hospitable. At the Bancroft Library, Mrs. Eleanor Bancroft, Mrs. Edna M. Parratt (now of the California Historical Society), and Mrs. Florence Kirschenbaum were very helpful. At the California State Library, a most friendly reception and invaluable aid were received from Miss Mabel R. Gillis, the state librarian, Miss Caroline

Wenzel, the head of the California Division, and Neal Harlow, now of the University of California at Los Angeles. At the Henry E. Huntington Library the writer had so pleasant and profitable a visit during 1941 that he repeated the experience during the summer just ended. There, to mention a few of the many from whom such willing assistance was received, thanks should be given to the librarian, Leslie E. Bliss, to Lyle H. Wright, Carey S. Bliss, Miss Mary Isabel Fry, Mrs. Geneva Johnson, Miss Josephine Baker, Miss Georgiana Patty, and Erwin Morkisch (who took the photographs from which the illustrations were made). In San Francisco use was made of the collections of the California Historical Society and the Society of Pioneers; in Sacramento many evenings were spent at the Public Library, where Allen Ottley (now of the state library) was especially helpful.

After research had been completed and the manuscript had begun to take shape, aid of a different kind was received. Robert Glass Cleland, of Occidental College and the Huntington Library, read the entire manuscript and made suggestions which have been incorporated into the final version. More recently, during the past summer, he has placed the writer still further in his debt. Anson S. Blake, formerly president of the California Historical Society, joined with his brother, Edwin T. Blake, a mining engineer, in reading the manuscript and providing criticisms which have saved the author from a number of errors. Louis B. Wright, of the Huntington Library, and Charles A. Barker, formerly of Stanford University, now of the Johns Hopkins University, read the manuscript and were generous in encouraging the author. Mrs. Marion Blackwell, of Pasadena, made the black and white sketch map which appears in this volume.

Here in the East frequent use has been made of the resources of the Harvard College Library, until recently the writer's home port, and of the American Antiquarian Society, the Massachusetts Historical Society, the Boston Public Library, and the Library of Congress.

Finally, there is a more personal debt to friends for help and company—particularly to Richard W. Leopold, Duncan S. Ballantine, and the group at 1619 34th Street. To the writer's family, both at Boston and Baltimore, the debt is even greater.

Rodman W. Paul

New Haven, Connecticut
September, 1946

CONTENTS

CONTENTS

ILLUSTRATIONS

All the above, save the sketch map of central and northern California, are reproduced through the courtesy of the Henry E. Huntington Library, San Marino, California. The sketch map was drawn by Marion Blackwell.

CALIFORNIA GOLD

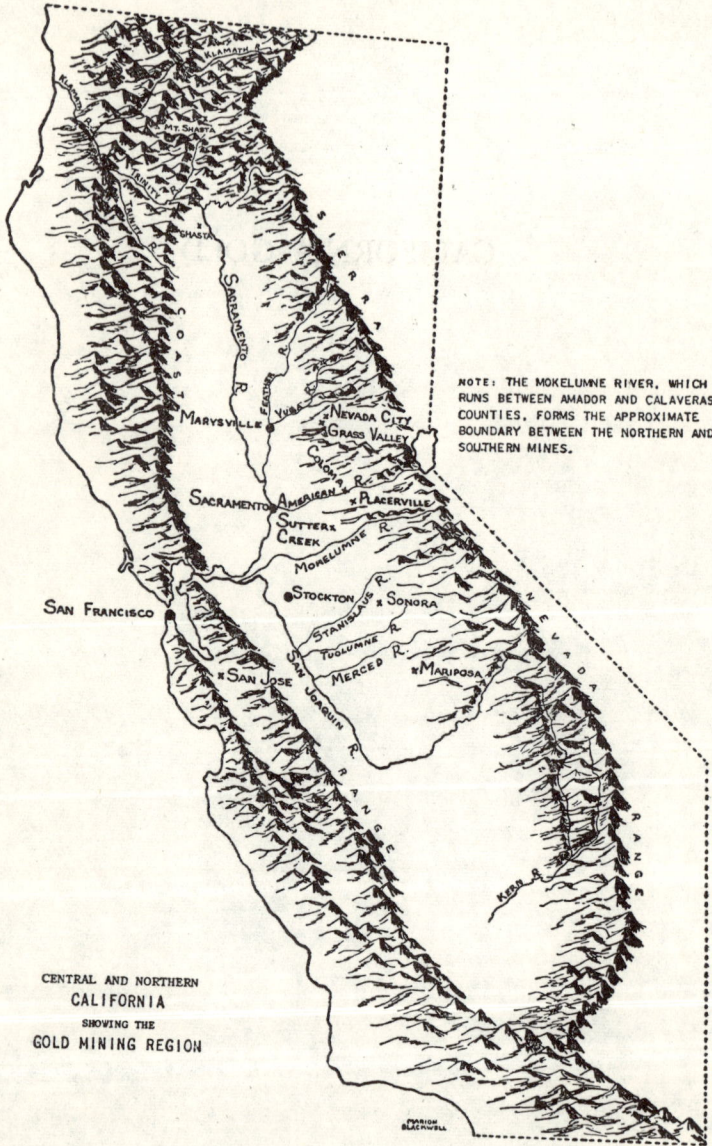

NOTE: THE MOKELUMNE RIVER, WHICH RUNS BETWEEN AMADOR AND CALAVERAS COUNTIES, FORMS THE APPROXIMATE BOUNDARY BETWEEN THE NORTHERN AND SOUTHERN MINES.

CENTRAL AND NORTHERN
CALIFORNIA
SHOWING THE
GOLD MINING REGION

THE SETTING

In the months of February and March, 1848, the United States Senate was the scene of one of those bitter political struggles which have so often attended a crisis in the nation's foreign policy. For two years the country had been at war with Mexico. To the surprise of the world, the raw young republic had defeated its southern neighbor in a succession of campaigns, and had overrun an imperial extent of its antagonist's territory. By the opening of 1848, Mexico had lost its capital city and seemed on the verge of losing all semblance of government and civil order.

Faced thus with disaster, the vanquished nation had agreed to terms of peace. A treaty had been drawn up and its provisions had been approved by Mexico. In February the treaty had reached Washington, and after winning the reluctant endorsement of the President, it had been submitted to the Senate.

It was this treaty which was occupying the attention of the Senate as the month of February drew to a close. Behind locked doors the upper house debated the problem in "long and continuous sittings" that were "fatiguing and oppressive." [1] For a time there was doubt

[1] Washington *National Intelligencer*, March 11, 1848.

that the necessary two-thirds majority could be mustered, and the uncertainty was the more understandable since no less a figure than the great Daniel Webster was leading the opposition. Webster proposed that the treaty be put aside so that a new agreement could be negotiated with the defeated adversary. But on this occasion the "god-like Daniel" failed. On March 2 his resolution was voted down, and on March 10 the treaty was ratified with a few votes to spare. A spiteful contemporary labeled it *"a Peace which every one will be glad of, but no one will be proud of."* [2]

With the benefit of hindsight, one may well wonder what course history would have taken if Webster's motion had prevailed. By the terms of the treaty Mexico agreed to surrender a spacious domain that included the present state of California. Unbeknown to anyone in Washington, on January 24 an American millwright, while erecting a small sawmill on the banks of a California stream, had picked up a bit of yellow metal that to his delighted amazement proved to be gold. News of his great discovery was not immediately announced, but by the time that the senators finally decided to approve the treaty, excited fortune hunters in California were already preparing to hurry to the site of that famous sawmill.

The discovery of gold, coming simultaneously with the transfer to American sovereignty, abruptly changed

[2] Washington *National Intelligencer*, March 14, 1848.

the direction of California's development. Until that time the territory had been a vast, sparsely settled colonial province inhabited by a large population of Indians and a small number of Caucasians. Neither race had done much to further California's progress. The Indians, despite being more numerous than in most parts of aboriginal North America, had distinguished themselves chiefly by the cultural poverty of their attempts at civilization. Not only had they failed to achieve any political organization worthy of the name, but they had not raised even the simplest manual arts above the absolute minimum required for existence.

The Spanish, who succeeded them as masters of California, had done better. They had visited California as far back as 1542, when Cabrillo, a Portuguese in the service of Spain, had directed his two small ships into the harbor which has since become known as San Diego. Colonization, on the other hand, had not begun until 1769, and it had never been supported with resources adequate for the task at hand. As late as 1820 the total white population was only a little more than three thousand.

This corporal's guard of a provincial population was not scattered over the face of California, but was rather gathered into groups, at the missions, presidios, and pueblos. All of these foundations were strung along the rim of the seacoast from San Francisco Bay to San Diego —a distance of five hundred miles; none was located much more than a day's ride from the sea. The Spanish were not unfamiliar with the interior, but they established no colonies there.

The missions were the core of Spanish California. They were more numerous and more important than either the presidios, which were the governmental and military centers, or the pueblos, which were nascent towns. In the missions, church and state joined hands to convert the aboriginal population from heathen idleness to Christian usefulness. Here was achieved whatever progress the colony made in rural industry. Under the padres' guidance the Christianized Indians produced grain in abundance, experimented with irrigation, and proved that orchard, vineyard, and garden crops could be grown. They proved also that California was ideally fitted by nature for the raising of cattle, sheep, and horses.

In demonstrating California's suitability for livestock, the missions were preparing the way for their own demise, for in 1822 Spanish rule in upper California was brought to an end, and in its place came a weak Mexican administration that soon fell under the dominance of land-hungry Mexicans and Californians who wished to turn California into a ranchers' paradise. The missions were broken up and their lands were parceled out among men whose sole interest was in raising cattle and horses. Vast, undeveloped ranches, marked only by a few crude adobe buildings, came to be the characteristic form of settlement. Breeding half-wild cattle came to be the only form of economic effort. When manufactured goods were needed, the Californians were content to slaughter their cattle for hides and tallow, and barter the resultant products for the supplies brought to the coast by ship-borne American and British merchants.

To estimate the relative significance of this pastoral economy one must realize how small was the population and how limited its activities when compared with the size of California and with the potentialities of the province's natural endowment. The state stretches along the seacoast for 750 miles, and it extends inland for nearly 200 miles. In shape it somewhat resembles a parallelogram, in which the sloping sides trend in a northwest-southeast direction. In surface topography it appears, when one first glances at a relief map, to consist chiefly of mountains, save for a single broad, flat strip that runs through the heart of the interior parallel to the seacoast. A closer examination shows that two-thirds of the mountains are knit into two great longitudinal systems: the Coast Range and the Sierra Nevada. In between the two lies the Great Valley of California,* that strip of level inland territory which is so prominent upon the relief map. Both at the north and the south the Sierras and the Coast Range curve together to lock in the Great Valley from access to the outer world.

At only one point is there a break in this surrounding

* Also known as the Great Central Valley or Central Valley.

wall of mountains. Two-thirds of the way up the California coast, as one goes from south to north, is the bay of San Francisco. At its northeast corner this bay is joined to the Great Valley by a chain which consists of two smaller bays—San Pablo and Suisun—and a narrow passageway which serves as the link between them: the Straits of Carquinez.

This waterway supplies the one resource that the Great Valley would otherwise lack: access to the outer world. In all other respects the region's natural advantages are boundless, since it has both agricultural capacity and mineral richness. The valley itself is over four hundred miles long and has an average width of perhaps forty miles. The northern half is drained by the Sacramento River and its several tributaries; the southern half is less fully tapped by the San Joaquin River and its auxiliaries. Along the edges of the valley's floor, foothills lead up steeply to the mountains. Those on the Sierra side have won permanent fame as the site of the original gold discovery. From them and from the ranges immediately above them have been extracted most of the precious metals that California has produced.

To the west of the Great Valley, on the seaward side of the Coast Range, are nine smaller valleys that have been of great importance to California's agricultural development, but have played only a limited part in the state's mining history. Because they are hidden in the mountains near the borders of the sea and along the edges of the bays, they are generally referred to as the coastal valleys. They appear on the relief map as longitudinal troughs that have a general trend parallel to the seacoast.

These coastal valleys and the Great Valley, the Sierra Nevada and the Coast Range mountains, and the waterway into the interior together constitute the heart of California. This is the section which dominated the history of the state during the quarter-century between 1848 and 1873. Here gold was first discovered, and here the most significant part of the mining industry developed. Here the first cities were founded, and here American agriculture began. Here, too, an adequate system of transportation arose for the first time in the Far West. Almost could one say that the story of this section was the history of California during the twenty-five years that followed the gold discovery.

The map, however, shows that this is but the central part of California. On all sides save the west it is hemmed in by land whose total extent, within the state, is approximately the equivalent of the central area itself. That this border country played a role which was comparatively slight during the period can be explained largely in terms of natural factors. Up in the northwest corner of the state the land is so broken by mountains and so handicapped by a severe climate that the advance of civilization has reached it only very slowly.[3] In the

[3] Cf. Owen C. Coy, *The Humboldt Bay Region, 1850-1875, A Study in the American Colonization of California* (Los Angeles, 1929).

northeast corner great beds of barren lava have robbed the countryside of its value to man. Throughout the narrow strip to the east of the Sierras' crest there has been the constant handicap of the mountain barrier formed by the Sierras themselves. This barrier deprives the eastern edge of the state of both rainfall and easy communications. Farther south, in the southeastern portion of the state, there are the obstacles of endless deserts and rainless, desolate mountains.

There remains the southwestern section—that area which is customarily termed southern California. During the years from 1848 to 1873 southern California was anything but a progressive region. When compared with the hustling American civilization that was arising in the heart of the state, it seemed a sleepy heritage from California's Spanish-Mexican past. It was held back by its remoteness from the new population centers to which the Gold Rush had given rise. It was retarded by the aridity of its climate, and it suffered from the lack of initiative shown by its predominantly Spanish-Mexican population.[4]

In all these parts of California, save perhaps the northwest, there was one omnipresent factor that had always to be taken into account, and that was the famous California climate. As all the world knows, California has a larger amount of sunshine and a smaller amount of rain

[4] Cf. Robert G. Cleland, *The Cattle on a Thousand Hills: Southern California, 1850-1870* (San Marino, California, 1941).

than the states east of the Mississippi. What is more important is that the rains fall only during the winter season —from late November to May. For six months there is blazing, burning sunshine, and for the other six months there is an incredible amount of rain, whole weeks of uninterrupted downpour and drizzle before one sees again the welcome sight of the temporary return of the sun.

This sharp division into a wet season and a dry one has been one of the decisive influences in the life of the state. That it should affect agriculture is obvious. What may not be so apparent is that it has to an important degree determined the characteristics of mining. In economic significance the climate deserves to rank with the wealth of the Sierra Nevada and the fertility of the valleys as a factor which helped shape the history of the state during the twenty-five-year period which began with the discovery of gold in 1848 and ended with the transition into a new era in 1873.

II

THE NEW AGE BEGINS

During the mission and ranching periods the Spanish Californian civilization had spread its influence over the strip of land contiguous to the ocean and over the rich valleys that led inland from the sea. From San Diego in the south to the northern shores of San Francisco and San Pablo bays could be found the scattered bits of evidence that proved the presence of civilized man: isolated ranches and herds of cattle, a few missions, a few towns, and a system of dusty trails that passed for roads.

Meanwhile the vast expanse of the Great Valley remained almost as untenanted by white men as it had been on the day when Cabrillo touched at San Diego in 1542. One American and a few Spanish Californians had begun ranching at widely separated points east of the Coast Range, but they had received little support from the rest of the California population.[1] The interior was disliked because of its remoteness from civilization and from the Pacific Ocean, the only artery of commerce then in use. The interior was feared because of the ferocious reputation of its Indians, who had never felt the gentling influence of the padres.

[1] George D. Lyman, *John Marsh, Pioneer, The Life Story of a Trailblazer on Six Frontiers* (New York, 1930), pp. 206-230.

Finally, in 1839, California was visited by a Munchausen-like character who had ambitions to become a feudal baron on the frontier. John A. Sutter was Swiss in nationality and magnificently impressive in personality. After a dubious career in several parts of the world, he was deposited on the shores of California. There, by the simple expedient of fabricating a glamorous European past, he secured both recommendations and credit.[2]

He gathered together a small expedition, loaded it into two little schooners and a four-oared boat, and sailed across the bays of San Francisco and San Pablo to the mouth of the Sacramento River. Up this he voyaged, amidst mosquitoes and hostile Indians, to the site of the present city of Sacramento. There he built his famous fort, New Helvetia, and there he developed the first important settlement by white men in the Great Valley. With a motley collection of half-tamed Indians, Hawaiian Islanders, and white outcasts, he managed to develop a crude system of agriculture, an equally crude grist mill and distillery, and a considerable amount of stock raising.

The success of Sutter's experiment led to a small "boom" in the Sacramento Valley during the early and middle forties. A Scotsman established a ranch two miles from New Helvetia, on the American River, a

[2] Of the several accounts, the most complete is James P. Zollinger's *Sutter: the Man and his Empire* (New York, 1939).

tributary of the Sacramento. Three Germans went farther north to invade the valleys of the Feather and Yuba rivers, two streams whose waters unite to flow into the Sacramento. A French and an American sailor "took up" land on Bear River, just south of the Yuba. A group of Americans, more than a dozen strong, pressed on towards the head of the Sacramento Valley and settled along the Upper Sacramento and its several northern tributaries. By the opening of 1846 isolated ranches were scattered through the Sacramento Valley all the way from the junction with the San Joaquin to the site of the present town of Red Bluff, where a transverse ridge of upland robs the valley of its head.

A similar development in the San Joaquin Valley was several years later in starting. Not until 1847 did Charles M. Weber, a courageous German-American, succeed in founding upon the site of the modern city of Stockton a colony that was to be for the San Joaquin Valley what Sutter's Fort was for the Sacramento: the hub from which other settlements radiated.

Beyond the western limits of the Great Valley, during this same period, Americans and Europeans were filtering into three of the rich coastal valleys tributary to San Francisco Bay—Santa Clara, Napa, and Sonoma—and were establishing themselves at scattered points around the rim of the bay itself. The village of San Francisco, sometimes called Yerba Buena, was fast becoming an American community. Down at the southern tip of San

Francisco Bay, at the town of San José, Anglo-Saxons were numerous enough to elect one of themselves to the chief municipal office, while to the south of San José, on the shore of central California's second great bay, Monterey, Americans were trying the prospects of the little settlement at Santa Cruz.[3]

By this occupying of isolated ranches and invading of established towns, Americans, and the Europeans associated with them, rapidly took possession of some of the most desirable locations in the area which forms the heart of California. Hispanic civilization had started in southern California and had crept northward along the Pacific shore and up the coastal valleys. American colonization was concentrated in the north, and had its inception in the Great Valley and the valleys tributary to San Francisco Bay. It was this part of the province, rather than the older section to the south, that was to be the central arena for California's history during the next quarter-century.

The number of American settlers engaged in this infiltration was less impressive than the geographical scope of their activities. The historian Hubert Howe Bancroft has estimated that in 1845 there were less than 700 men in California who were not of Spanish blood, out of a total population (exclusive of Indians) that was more than ten times that figure. After 1845 the ratio shifted

[3] On the spread of American settlement, see Hubert H. Bancroft, *Works* (San Francisco, 1882-1890), XXIII, 4-21.

markedly as large numbers of Americans came into the province. According to Bancroft's estimate for the middle of 1848, just as the Gold Rush began, the total population was then 14,000, of whom 7,500 were Spanish Californians and 6,500 were "foreigners." [4]

What sort of a civilization these newcomers would have evolved, one can only guess. Very likely it would at first have been little more than an Americanized version of the Hispanic cattle-raising, horse-riding existence. The lack of transportation facilities and the limited nature of the local market would have long retarded agriculture in the interior. Another deterrent would have been the scarcity of lumber. California as a whole is not well wooded. The Sierras and the Coast Range are heavily timbered, but the floor of the Great Valley is a vast plain devoid of trees, save for a few oaks and a little timber along the rivers' edge. From the earliest days of his colony, Sutter had been handicapped by this lack. Finally, in August 1847, he decided that by setting up a sawmill in the Sierras' foothills he could simultaneously assure himself of a supply of building materials and obtain a good income by selling planks and beams to immigrants. He therefore contracted with James W. Marshall, an eccentric millwright and carpenter, for the construction of a mill on the American River, forty miles' ride from New Helvetia.

By the opening of 1848 this mill was nearly ready for

[4] Bancroft, *Works*, XXXII, 524, 643, XXIII, 3, 158-159, n. 23.

operation. It lacked but one thing for completion, and that was a deeper tailrace. In order to secure one without extensive labor, Marshall built a diversion dam, by means of which the river was to be compelled to cut its own way. On January 24, 1848, one of his helpers noted in his diary: "This day some kind of mettle was found in the tail race that looks like goald, first discovered by James Martial, the Boss of the Mill." [5]

Because he was something of a visionary, Marshall at once jumped to the conclusion that he had indeed discovered the precious metal. His men were more skeptical. They tested the yellow flakes and grains by every method their minds could suggest. They bit the "goald," they pounded it with a hammer, they heated it in a fire, and they plunged it into boiling lye. When they had finished with it, Marshall rode down to New Helvetia, and, behind closed doors and in an excited whisper, he described the discovery to Sutter. Sutter, too, was dubious, and the testing began again. Sutter took some nitric acid from his medicine chest and tried it on the metal. Then he got a scales and weighed it. Finally he hunted up the *Encyclopedia Americana* and studied carefully the long article on gold. [6]

By the time that Sutter was convinced, the discoverer

[5] Facsimile in *Pictorial History of California*, ed. by Owen C. Coy (Berkeley, California, 1925), p. 119.

[6] In addition to Zollinger, *Sutter*, pp. 236-237, and Bancroft, *Works*, XXIII, 33-38, see "The Discovery of Gold in California," *Hutchings' Illustrated California Magazine*, II (1857-58), 194-195, printing Sutter's and Marshall's statements.

was so uncontrollably excited that, as Sutter later re-called it, "Marshall wanted me to return with him to Coloma [the site of the mill] at once, but it was nearly supper time and pouring rain, so I said I would go the next morning. Yet Marshall was so impatient that he started back that evening, in the rain, with no supper." [7] The gold fever had caught its first victim.

Sutter rode up to Coloma the following day. He looked at the millrace, and then had a talk with the workmen. He had several important building projects in progress, and was anxious to complete them. Realizing that news of the discovery would cause a stampede, he asked the men to keep the great news secret for six weeks. They agreed, but Sutter forgot to silence the gossiping tongue of the feminine cook at the mill. She and her young sons soon betrayed the secret, and Sutter's own indiscretions added to the leakage.

Thanks to the paucity and slowness of communications, and to men's skepticism, a real "rush" did not develop for over a month. Sutter's employees at the fort started deserting early in March, but the first reports of the discovery did not appear in the two little newspapers then published at San Francisco until March 15 and March 25, respectively. During April the excitement increased, and in May it became uncontrollable. On May 29, 1848, the San Francisco *Californian* suspended publication with the bitter explanation that so normal an

[7] *Ibid.*, p. 195; obviously a paraphrase rather than Sutter's own English.

institution as a newspaper had become entirely super-
fluous. "The whole country, from San Francisco to Los
Angeles, and from the sea shore to the base of the Sierra
Nevadas, resounds with the sordid cry of 'gold, GOLD,
GOLD!' while the field is left half planted, the house
half built, and everything neglected but the manufacture
of shovels and pickaxes." [8]

[8] As quoted in Edward C. Kemble, "The History of California Newspapers," *Sacramento Daily Union*, December 25, 1858.

III

THE GREAT MIGRATION

FORTY-EIGHT, FORTY-NINE, AND THE EARLY FIFTIES

In the history of almost any of the American states one can find a few events and trends that in significance stand out from the purely local as sharply as a single tree upon a desert plain. The Gold Rush is an instance. Within California the Gold Rush was a revolution that changed forever the character of the state. Beyond California's boundaries it was a magnetic force that sent its lines of attraction into every nation.

Within two years of Marshall's discovery every civilized country knew the name "California," although at that time few persons in Europe, or Asia, or Latin America could have identified Illinois, or Massachusetts, or Georgia. In all parts of the world men made ready to seek the Golden Fleece, or to venture their money in the expeditions organized by those more daring than themselves, or to invest in the expansion of trade and shipping to which the Gold Rush was giving rise.

Even if cautious men decided to remain at home and keep their capital with them, still they could not escape entirely the effects of the train of events set in motion at Coloma. As has often happened in history, the discovery in one area soon inspired like discoveries in other regions, notably in Australia in this instance. Presently the basic medium in the world's currency began to increase, more rapidly than in any previous period of which there is record. In the United States alone, gold production multiplied seventy-three times during the six years that began in 1848, and from a position of insignificance among the gold producing nations, the United States climbed upward so rapidly that during the period from 1851 to 1855 it contributed nearly 45 per cent of the world's total output. The result was an inflation that affected all the countries on the earth's surface.[1]

More dramatic and more immediately obvious was the functioning of the Gold Rush as a population movement. Here the first phase was confined to California itself. It consisted of a tumultuous stampede to the Sierra foothills by the people already in the province. The exodus began from Sutter's Fort in March and April, from San Francisco and Sonoma in May, from San José, Monterey, and Santa Cruz in May and June, and, finally, from the strongly Hispanic southern communities in July and August. As the chronology suggests, the excitement started in the American and European strongholds in the north and the interior, and worked gradually southward into the region settled primarily by Spanish Californians. The Americans, and the Europeans associated with them, thus became at the very beginning the

[1] Robert H. Ridgway and others, "Summarized Data of Gold Production," U. S. Bureau of Mines, *Economic Paper*, no. 6 (Washington, 1929), pp. 14, 61.

dominant power in the mines. Nor did they ever thereafter let control slip from their hands.[2]

Until midsummer, 1848, California's own population had sole possession of the mines. Then commenced the rush from outside the province. In June the great news reached the Hawaiian Islands, and in July crowded vessels began ferrying gold seekers—white, Kanaka, and half-breed—across the Pacific. In early August Oregon heard of the discovery and began sending delegations southward. Mexico learned of it somewhat later, and apparently did not contribute until October. Peru and Chile knew of it as early as September, and sent several thousand men northward in the closing months of 1848 and the early part of 1849.[3]

Rumors of the discovery reached the Atlantic seaboard of the United States in August, but did not attract much attention until mid-September, and they were not established as fact until confirmed by the President's message to Congress on December 5.[4] For that reason no Americans save those from Oregon and Hawaii reached California during 1848. The first to arrive were those who sailed into San Francisco harbor on February 28,

[2] Bancroft, *Works*, XXIII, 71, 83; Jacques A. Moerenhout, *The Inside Story of the Gold Rush*, trans. by Abraham P. Nasatir, California Historical Society, Special Publications, no. 8 (San Francisco, 1935), pp. 2-3.

[3] Bancroft, *Works*, XXIII, 111-113, 125; Moerenhout, *Inside Story*, pp. 41-46; Doris M. Wright, "The Making of Cosmopolitan California, An Analysis of Immigration: 1848-1870," *California Historical Society Quarterly*, XIX (1940), 326-327.

[4] *New York Herald*, August 19, September 14, 17, 1848; Bancroft, *Works*, XXIII, 114-115.

1849, at the conclusion of the voyage which inaugurated steam communication between New York and the Pacific Coast, via the Isthmus of Panama.[5]

The bulk of the American immigration did not reach California until the vanguard of the Cape Horn sailing vessels had made port in June, July, and August, 1849, or until the overland caravans had begun to straggle in during August and September. Direct immigration from Europe was even later, and can hardly be placed much before the close of the year. Indirectly, a considerable number of Europeans of dubious background came to California during 1849 and 1850 from England's Australian penal colony.[6]

How many joined in the rush of "forty-eight," "forty-nine," and the early fifties will never be known with any degree of accuracy, but it is clear that the influx of 1848 was comparatively so small that it was no more than a prelude to the stampede of 1849 and the subsequent years. Apparently the number of persons in the province, other than Indians, increased from about 14,000 in the summer of 1848 to somewhat less than 20,000 at the end of the year. At the latter time there

---

[5] John H. Kemble, *The Panama Route, 1848-1869*, University of California Publications in History, XXIX (Berkeley and Los Angeles, 1943), 35-36.

[6] T. Butler King, "T. Butler King's Report on California. Message from the President," *House Exec.* Doc., 31 Cong., 1 sess., no. 59 (March 27, 1850), p. 26; Rufus K. Wyllys, "The French of California and Sonora," *Pacific Historical Review*, I (1932), 338-339; San Francisco *Weekly Alta California*, November 23, 1850, March 1, 1851.

must have been between 6,000 and 8,000 Americans, and several thousand Latin Americans, Europeans, and Pacific Islanders, in addition to the Spanish Californians.

Twelve months later, at the close of 1849, the population, exclusive of Indians, was probably several thousand short of 100,000, with Americans forming between one-half and two-thirds of the total. When the United States census was taken in the middle of 1850, amidst turbulent conditions that made accuracy impossible, it returned a total of only 92,597, but this figure was at least 19,000 less than the true number, since the returns were lost for San Francisco and for two other counties.[7]

Because of the imperfect nature of the United States census of 1850, the state of California took a census of its own two years later. By that time the state had felt the full force of the Gold Rush. A host of eager gold seekers had come hurrying to El Dorado over the routes pioneered by the forty-eighters and forty-niners. There had been losses along with the gains, for some of the very dissatisfied and the very successful had soon left California, and the death-rate had been very high. The state had, however, doubled its population. Its total in 1852

[7] On population, 1848-1850, see: *Seventh Census, 1850*, p. 969; "Memorial," in J. Ross Browne, reporter, *Report of the Debates in the Convention of California, on the Formation of the State Constitution, in September and October, 1849* (Washington, 1850), pp. xxii-xxiii; "Population of California," *American Quarterly Register and Magazine*, III (1849), 385-386; Bancroft, *Works*, XXIII, 158-159; San Francisco *Weekly Alta California*, April 26, 1851; *Sacramento Transcript*, May 4, 1850.

seems to have been 223,856.[8] One may summarize these several sets of figures by saying that California's population jumped from 14,000 in 1848 to something less than 100,000 at the close of 1849, and that it then advanced to 223,000 by the latter part of 1852.

This abrupt increase was by no means caused solely by American immigration. The 1850 census showed that only 68 per cent of the 92,597 recorded had been born on soil that was United States territory prior to the Mexican War. A little more than 24 per cent had been born in "foreign countries," and that figure did not include the Spanish Californians. By contrast, in the United States as a whole in 1850 only 10 per cent had been born abroad. Wisconsin alone, of the other states, had a larger percentage of foreign born.[9]

Within the 24 per cent from "foreign countries," Latin Americans formed both the largest and most noticeable element. In this California was unique. No other part of the United States, up to this time, had been subject to Latin American immigration while under the Stars and Stripes. Chileans and Peruvians, several thousand strong, came northward by sea during the autumn of 1848 and the winter and spring of 1849. Because the

[8] The state census was carelessly done. For it see "Governor's Message; and Report of the Secretary of State on the Census of 1852," California Senate *Document*, 4 sess., 1853, no. 14 (January 26, 1853); and *Seventh Census, 1850*, p. 982.

[9] James D. B. De Bow, comp., *Statistical View of the United States, Embracing its Territory, Population . . . Being a Compendium of the Seventh Census* (Washington, 1854), pp. 61, 65, 118.

first-comers were treated in an unfriendly manner by the Anglo-Saxons, and at times were subjected to outright persecution, immigration from those two nations slacked off after 1849 and ceased entirely after the early fifties.[10]

The Mexicans suffered a like fate. They generally traveled to California by land, over an old Spanish colonial trail. They came variously from the northern provinces of Sonora, Sinaloa, Chihuahua, and Durango, but the predominance of men from Sonora was noticeable enough to establish the name "Sonorans" as a term universally used by Californians to describe the whole group. "Some of these Sonorans were intelligent men who came on their own responsibility, but the greater majority belonged to the peon class and were fitted out or grub-staked—to use a mining term—by their patrons who were to receive a share of their profits."[11] In two respects the Sonorans were unlike the other immigrants of the period. In the first place, they did not make a single journey to the land of gold, but instead performed an annual folk migration in which several thousand of them came north each spring and returned home each autumn. In the second place, a fair proportion brought their women and children with them. It was true of almost all the other gold seekers, both American and foreign, that when the men started for El Dorado, their

[10] Wright, "Cosmopolitan California," pp. 326-327.
[11] James M. Guinn, "The Sonoran Migration," Historical Society of Southern California, *Annual Publications*, VIII (1909-1911), 31-32.

families remained behind. If the census of 1850 is to be believed, females formed only a little over 7½ per cent of the recorded population in the census year.[12]

The movement northward from Mexico began in the latter part of 1848 and reached its maximum in 1850. Thereafter it decreased, until by 1854 it had ceased almost entirely. As was true of the Chileans and Peruvians, the hostile attitude encountered in California had much to do with bringing it to an end.[13]

A full year elapsed after the arrival of the first Latin Americans before the Europeans began making port at San Francisco. Among them, immigrants from the British Isles formed the largest group, Germanic peoples the second largest, and the French the third. The combined numbers of the English, Scotch, Irish, and Welsh were more than twice those of the Germans, Austrians, and Dutch.[14]

Both the English-speaking and German-speaking Argonauts seem to have fitted into the California scene with a minimum of friction, and both made notable contributions to the development of the state. The French did not escape so easily. For them, as for the Germans, the revolution of 1848 and the attendant economic and political disorder in Europe created a situation naturally conducive to emigration. To this was added the artificial stimulus of promotional work by companies that stood

[12] *Ibid.; Seventh Census, 1850,* p. 972.
[13] Guinn, "Sonoran Migration," p. 33.
[14] De Bow, *Statistical View,* pp. 117-118.

to profit from an exodus to California. One of the companies, for example, held a lottery, known as the Golden Ingot, which encouraged several thousand penniless Frenchmen to come to California during 1852.

Unlike the Germans, the French tended to keep to themselves, in groups of their own kind. By so doing they aroused the same sort of antagonism that had already caused difficulties for the Latin Americans. The Anglo-Americans were apt to refer to them irritably as "Keskydees," because whenever an English-speaking person tried to address them through an interpreter, the Frenchmen kept interrupting excitedly: "Qu'est-ce qu'il dit?" [15]

The hostility with which the French and Latin Americans were greeted was minor compared with the attitude towards the Chinese. China, like Latin America, was for the United States an entirely new source of immigration. The census of 1850 listed only 660 from that ancient civilization, but during the next few years the rate of influx increased so rapidly that one cannot lightly reject the claim of the 1852 state census that there were then 25,000 in the state. Their coming raised a serious problem in race relations. The Chinese differed sharply from all other immigrants in appearance, habits, and standards. It is little to be wondered that the year 1852

[15] J. D. Borthwick, *Three Years in California* (Edinburgh and London, 1857), pp. 241-242; San Francisco *Daily Alta California*, February 10, May 5, May 30, June 3, June 28, 1852, January 14, 1854; Wyllys, "French of California," pp. 337-342.

was less than six months old before the Chinese became the object of the first of many political agitations.[16]

All of these races—Latin American, European, and Asiatic—were subordinate in numbers and in significance to the new ruling element: the Americans. According to the census of 1850 the Americans composed two-thirds of the total population, and that ratio just about expresses the importance of their contribution to the building of the state during its early years.

The Americans were by no means a homogeneous group. On the contrary, they were as varied in their origins and occupations as the nation to which they belonged. Men of all trades and all sections had been caught by the magic of the word *gold*—gold that could be had for the asking, gold that could be scraped up in shovelfuls, gold that would make a man rich for life. From hard toil with the plow and the hoe, from the monotony of account books, from the grim reality of the seaman's routine, from the struggle on the western frontier—from all these the Argosy of the nineteenth century offered relief. And to those who had already gone restlessly into the Mexican War, it offered a new outlet for unquiet energies. At the opening of 1849 the *New York Herald* declared:

The excitement relative to the gold mines of California continues with unabated fervor. It is daily fed with all sorts of

[16] Rodman W. Paul, "The Origin of the Chinese Issue in California," *Mississippi Valley Historical Review*, XXV (1938-39), 181-196.

reports. Every statement is caught up and swallowed with the greatest avidity. . . . At this moment the spirit of emigration seems to prevail mostly in the agricultural and commercial States of the North, as well as in some of the Southwest. In every Atlantic seaport, vessels are being fitted up, societies are being formed, husbands are preparing to leave their wives, sons are parting with their mothers, and bachelors are abandoning their comforts; all are rushing head over heels towards the El Dorado on the Pacific—that wonderful California, which sets the public mind almost on the highway to insanity. . . . Every day, men of property and means are advertising their possessions for sale, in order to furnish them with means to reach that golden land. Every little city and town beyond the great seaports, or within their reach, is forming societies either to cross the isthmus or to double Cape Horn.[17]

A Scotch traveler who arrived by the Panama route in 1851 reported: "Among the Americans *en route* for California were men of all classes—professional men, merchants, labourers, sailors, farmers, mechanics, and numbers of long gaunt Western men, with rifles as long as themselves." [18]

The service via Panama was the quickest way by which Americans could reach California. Argonauts could take the steamer from New York to the Isthmus of Panama, cross to the Pacific, and reach San Francisco in a total of thirty-three to thirty-five days. In return they would have to pay what was considered a high price. In December 1849, the quoted rates from New

[17] New York *California Herald-Extra*, January 16, 1849. A special edition of the *New York Herald*.
[18] Borthwick, *Three Years*, p. 32.

York to San Francisco were $380 first class and $200 steerage, with a fee of at least $30 more for crossing the Isthmus. The actual cost to the traveler was often much higher because of the extortions of ticket speculators and because of additional expenses incurred en route. The appearance of competition subsequently forced a reduction in the charge, especially for steerage passage, and from 1851 to 1856 the traveler could choose between the steamship line operating on the Panama route and Commodore Vanderbilt's service via the Isthmus of Nicaragua.[19]

Even after he had secured a ticket the Argonaut still faced the obstacles of dangerously overcrowded steamers, unexpected delays that might cost him days or even weeks, and a fever-ridden journey across the strip of land that divided the Atlantic from the Pacific. Prior to the completion of the Panama railroad in 1855, it took three or four days to cross Panama. The voyager had to travel first up the fever-infested Chagres River, in native dugouts, and thence overland on muleback. Often this terrible crossing was made in a tropical downpour, through clouds of mosquitoes and ankle-deep mud, and with only the most primitive of overnight accommodations. It is little to be wondered that many a gold seeker ended his quest forever on the Isthmus of Panama.[20]

[19] Kemble, *Panama Route*, pp. 54-94, 148, 166-173.
[20] Borthwick, *Three Years*, pp. 10-25; Isaac Read, "The Chagres River Route to California in 1851," ed. by Georgia W. Read, *California Historical Society Quarterly*, VIII (1929), 3-16.

If an Argonaut could not afford the Panama steamer, he had a choice between four other roads to El Dorado. One possibility was to embark on a ship bound for the Gulf Coast of Mexico, ride across Mexico to the Pacific, and there resume the journey by water. An alternative that avoided the sea-borne legs of the trip was to start from Texas, Arkansas, or Missouri and travel over one of the several trails that led variously through northern Mexico and the American Southwest to California.

Only a minority of the Argonauts selected either of those two routes. The majority, if they could not go by Panama, chose either the long overland journey across the Great Plains and the Rockies—through the heart of America—or else followed the seagoing traditions of New England and the Middle Atlantic states by taking passage on a sailing vessel bound round the Horn. While the latter approach to El Dorado was relatively safe, the trip was apt to be a long and wearisome one in which the effects of bad food, cramped quarters, and inactivity combined to bring the would-be miner into San Francisco harbor in poor physical condition. In 1849 the time required for the Cape Horn voyage ranged all the way from four months to eight, with the average said to be about five and a half. During the following year the clipper ship *Sea Witch* proved that it was possible to reach San Francisco in ninety-seven days from New York, and in 1851 the clipper *Flying Cloud* cut a week from the record made by the *Sea Witch*, but the average vessel continued to require two months longer than those fleet craft.[21]

Far more demanding and yet, to many Americans, more appealing was the last of the possible roads to El Dorado, that of overland migration across the plains. In many respects that was the hardest journey of all. It meant spending the greater part of the spring and summer plodding across the infinity of the treeless Great Plains, with the dust swirling up in choking clouds from the feet of the oxen and mules, and with the water holes threatening a drought and the prairie grass a famine. It meant possessing the courage to face Indian attacks and the moral strength to survive the burning hell of alkali and salt lands in Utah and Nevada. Then, finally, it meant summoning up enough reserve stamina to surmount the Sierras at a time when one's own physical exhaustion was rivaled only by the spent condition of his oxen and mules.[22]

Like the sea voyage round the Horn, the journey across the plains varied in length and difficulty according to the particular circumstances of the individual party.

[21] Octavius T. Howe, *Argonauts of '49: History and Adventures of the Emigrant Companies from Massachusetts, 1849-1850* (Cambridge, Massachusetts, 1923), pp. 159-178, 181; San Francisco *Alta California* (steamer ed.), August 2, 1849, January 1, 1851.

[22] Three of the best accounts are: William Kelly, *An Excursion To California over the Prairie, Rocky Mountains, and Great Sierra Nevada* (London, 1851); Eleazer S. Ingalls, *Journal of a Trip to California, by the Overland Route across the Plains in 1850-51* (Waukegan, Illinois, 1852); Sarah Royce, *A Frontier Lady: Recollections of the Gold Rush and Early California*, ed. by Ralph H. Gabriel (New Haven, Connecticut, 1932).

William Kelly's wagon train made comparatively quick time. In March and April 1849, the members of his group were moving westward to Independence, Missouri. On April 16 they started across the prairie, and on July 26 they were in the California "diggings." Judge Ingalls' company started from Lake County, Illinois, on March 27, 1850, and reached Placerville, California, on August 21. Mrs. Royce's family headed across the level prairies of Iowa on the last day of April 1849, and barely surmounted the Sierras before the first snowfall.

Whether the nineteenth-century Argonaut traveled by land or by sea, of one result he could be certain: he would arrive in El Dorado exhausted in body and depleted in pocketbook; in short, in anything but favorable condition for beginning the difficult profession of mining. The drain upon the individual's resources during the journey was vividly illustrated by Judge Ingalls' wayside notation in his diary:

> The appearance of the emigrants has sadly changed since we started. Then they were full of life and animation, and the road was enlivened with the song of "I am going to California with my tin pan on my knee." "Oh, California, that's the land for me," but now they crawl along hungry, and spiritless, and if a song is raised at all, it is, "Oh carry me back to Old Virginia, to Old Virginia's shore." [23]

Precisely because it was so severe a test, the journey to California was something that the gold seeker never

[23] Ingalls, *Journal of a Trip*, p. 38. The entry is dated Humboldt Desert, July 28, 1850.

forgot—if he survived it. Again and again the hardships and wonders of the trip were narrated about camp fires in the Sierras' foothills. For many of the Argonauts it was their first excursion beyond the limits of their local community. Few raw country boys, or inexperienced clerks, or untutored westerners ever forgot their first sight of the lush tropical jungle of Panama, or their first day ashore in a foreign land at Rio de Janeiro, or "the solemn stillness of uninvaded nature, the measureless immensity of the regions around us" on the Great Plains. For all who shared in it, the expedition to California ranked with El Dorado itself as one of the formative experiences of their lives.

IV

FUNDAMENTALS AND TEACHERS

Writers have often commented upon the strange stroke of fortune that gave to Americans the honor of discovering gold in California. Historically the Spanish have been one of the world's great mining peoples, while for Americans, prior to 1848, searching for mineral wealth was an unusual occupation. Then, too, in 1848 Cabrillo's voyage up the Pacific Coast was three hundred and six years distant, and the Spanish communities at San Diego and Monterey were seventy-nine years old, whereas distinctively American settlements had existed for only a few years.

Yet there is no mystery about the failure of the Hispano-Californians to uncover the hidden treasure. The gold was secreted in the interior of the province: precisely the region that the Spanish race had not colonized. Had the mineral deposits been located near the sea, they would doubtless have been uncovered much earlier, perhaps by the first civilized settlers. Indeed, the Spanish had discovered gold in southern California prior to 1848, and small amounts of the precious metal had formed a part of the return cargoes of some of the "Boston ships" that came to trade on the California coast, but neither the quality nor the quantity of the deposits had been sufficient to warrant a bona fide "rush." [1]

It was the occupation of the Sacramento Valley by Sutter that led to the discovery, and the site of Sutter's sawmill became the location for California's first mining camp: Coloma. Coloma, in turn, became the parent of the other early camps and "diggings."

To Coloma, in March 1848, came a Frenchman who had been cutting wood for Sutter at a place ten miles to the southeast. No sooner had he observed the characteristic features of the auriferous countryside surrounding the mill, than he was struck by the resemblance of that area to the one in which he had just been working. He hurried back to the scene of his lumbering operations and opened the successful gold "diggings" later known as Weberville.

A short time afterwards one of the American ranchers from the Feather River came down to the mill, was likewise impressed by the similarity to his own district, and hastened northward to pioneer the great Feather River gold fields, seventy-five miles from Coloma. A second American rancher, who had been living still further to the north, near the head of the Sacramento Valley, also made a pilgrimage to Coloma, only to realize that he, too, should have begun work at home. He

[1] Alfred Robinson, *Life in California, During a Residence of Several Years in that Territory* (New York, 1846), p. 190; Richard H. Dana, Jr., *Two Years Before the Mast, A Personal Narrative of Life at Sea* (ed. of 1868), p. 324, fn. (not in 1840 ed.).

turned back and discovered the Shasta diggings, two hundred miles northwest of Coloma.

Down in the San Joaquin Valley, Charles M. Weber, the German-American pioneer, decided to "prospect" his own region. He organized a party to search for minerals on the Stanislaus and Mokelumne rivers, two streams which plunge down from the Sierras to join the San Joaquin River. He found gold at once along the Mokelumne, and some Indians in his employ soon discovered deposits on the Stanislaus.[2]

Each of the diggings thus revealed became a center from which exploring parties went forth to make discoveries. Ranchers and sailors, hunters and farmers, all suddenly metamorphosed into miners, scrambled about in defiance of difficult topography and dangerous Indians as they excitedly sought quick fortunes. Like those of Coloma itself, most of the deposits unearthed were in the foothills which buttress the western flank of the Sierras. Some, however, were in the Shasta area, at the head of the Sacramento Valley, where the Sierras and the Coast Range curve together.

Before the close of 1848 miners were at work in diggings scattered along the western slope of the Sierras from the Feather River in the north to the Tuolumne River in the south, a distance of one hundred and fifty

[2] On the first diggings, see John S. Hittell, *Mining in the Pacific States of North America* (San Francisco, 1861), pp. 10-11; George H. Tinkham, *A History of Stockton from its Organization up to the Present Time* (San Francisco, 1880), pp. 71-73.

miles. Simultaneously, they were turning up the earth at numerous points within a region fifteen miles square at the Shasta gold fields. In the following year they pressed on beyond the Tuolumne to John C. Frémont's Mariposa estate, forty miles to the south.[3]

All of this activity was on the eastern and northern limits of the Great Valley region. Soon after opening these main fields, however, men thrust their way across the Coast Range from the head of the Sacramento Valley, and began exploiting extensive deposits in the rugged, remote section that forms the northwestern corner of California.[4]

By that time the auriferous areas were sufficiently well known so that men could begin to take stock of their findings. It was obvious that the most important gold-bearing ground was that found along the Sierras' western flank, and that throughout the greater part of its length this Sierran "field" had certain uniform characteristics that marked it as a single unit. To this unit the miners gave the name "Mother Lode."[5]

The Mother Lode occurs in a belt, or strip, of countryside that extends along the western slope parallel to

[3] Hittell, *Mining in the Pacific States*, pp. 17, 20.
[4] *Ibid.*; San Francisco *Alta California* (steamer ed.), August 2, 1849.
[5] The following discussion of the Mother Lode and of gold deposits is based on: Adolph Knopf, "The Mother Lode System of California," U. S. Geological Survey, *Professional Paper*, no. 157 (Washington, 1929); Clarence A. Logan, "Mother Lode Gold Belt of California," California State Div. of Mines, *Bulletin*, no. 108 (November, 1934); Waldemar Lindgren, *Mineral Deposits* (4th ed.; New York, 1933), especially pp. 155-156, 217-230, 547.

the main axis of the mountain system. The belt is about 120 miles long and varies in width from a few hundred feet to two miles. Its northern extremity is a few miles north of Coloma; its southern termination is in Mariposa County, in the neighborhood of the Frémont estate. At its northern end, the Mother Lode belt is found at an elevation of about 2,700 feet; at its southern, at about 2,000 feet. In between the two extremes it dips down nearly to the 1,000 foot level. It thus occurs in "a hilly country of moderate relief." [6]

The nature of the Mother Lode can best be explained in terms of its geologic history. The lode was brought into being by the forces that produced the present Sierra Nevada range. Long ages ago a mighty earth movement known as the Cordilleran revolution snapped the earth's surface for several hundred miles along the eastern edge of California, and thrust up a great tilted block of the earth's crust. During this process the existing beds of rock were folded and crowded together. Then great masses of molten magma invaded the earth's crust, to form the granitic rocks that make up the present core of the range. During and shortly after this period of igneous activity, ore deposits of many kinds were formed, including gold, tungsten, and copper. These deposits were in the form of veins: broad ribbons of ore that flowed into fissures in the rock.

Since the Cordilleran revolution erosion has been at

[6] Knopf, "Mother Lode," p. 1.

work. It has removed a thickness of several thousand feet from the original Sierra Nevada, thereby reducing the gold area to its present topography. Included in the material eroded away were several thousand feet of the upper portions of the gold veins.

Prior to this erosion, but one type of gold deposit had existed: gold in solid veins firmly embedded in surrounding rock. By the eroding process, vein gold was torn away from the surrounding rock and was carried off by streams and rivers. In the streams and rivers the particles of gold were abraded and reduced in size by the grinding action of the debris carried by the flowing water. As the streams rushed downward, the heavier particles, both of gold and of debris, gradually fell to the bottom and lodged between stones, or in cracks. The finer particles were carried somewhat further before suffering a similar fate.

In this manner were formed gold deposits of a new type: fragments that were to be found along the channels of present or former rivers, intermixed with sand, gravel, and stones. The fragments varied in size all the way from fine dust up to masses weighing one hundred pounds. Miners have long classified this type of gold under a word of Spanish origin: *placer* gold. The term "Mother Lode country" has been used commonly to describe the entire area in which are found both the "mother vein," which is still embedded in the rocks, and the placer gold that has been eroded from it.

By a process similar to this, all of the gold deposits in California have been formed, both in the Mother Lode country and elsewhere. In every case, a period of earth movement has ended in the deposition of mineral ores which have been attacked subsequently by erosion. To-day the gold is sometimes found still in the vein, from which it can be extracted only by the costly method of crushing the surrounding rock. Sometimes it is found in the more accessible form of placers, where it needs only to be picked out from the loose material in which it is embedded.

Because of its uniquely great extent and the romantic glamor associated with it, the Mother Lode has tended to monopolize the attention of both writers and tourists. Actually it embraces but a part of California's auriferous area. Some of the richest and most enduring mines in the state lie outside the Mother Lode region, and there are literally hundreds of square miles in other parts of the state that contain significant amounts of gold.

The Mother Lode ends a short distance north of Coloma, which is in El Dorado County. At El Dorado County's northern boundary there begins a series of famous mining counties that are placed one on top of another like bricks in a wall. Each of these contains many well mineralized districts. First comes Placer, then Nevada, then Yuba and Sierra, then Butte and Plumas, and finally Shasta, the northernmost mining area within the wall which encloses the Great Valley. Across the wall lies the wilderness of the northwest, which includes the mining counties of Trinity and Humboldt, Del Norte and Siskiyou. Of all these, Nevada has proved to be the richest, and for the purposes of history it has been the most important mining county in the entire state.

The miners who were in the Golden State in 1848, 1849, and the fifties could not begin to exploit fully the vast potentialities of this mineral wealth. They were neither sufficiently numerous nor sufficiently experienced. A careful contemporary estimated that at the end of 1848 there were but 5,000 miners actually at work, that a year later there were 40,000, and that towards the end of 1850 there were 50,000.[7] During 1852 the number was said to be 100,000, and in the opinion of one observer that figure and that year marked the peak in the size of the mining population.[8] A leading newspaper claimed, however, that in 1855 there were 120,000 miners, of whom 20,000 were Chinese. At as late a date as 1861 a usually reliable authority set the total as high as 100,000.[9]

Another contemporary pointed out, in 1855, that the gold fields really began in southern Oregon and extended

[7] Hittell, *Mining in the Pacific States*, p. 20.
[8] John B. Trask, "Report on the Geology of the Coast Mountains; . . . Also, Portions of the Middle and Northern Mining Districts," California Senate, *Document*, 6 sess., 1855, no. 14, p. 80.
[9] *Sacramento Daily Union*, October 10, 1855; Hittell, *Mining in the Pacific States*, pp. 20-21; but cf. an estimate of 75,000-100,000 in March 1858, in "Tax upon Mining Claims," *California Mining Journal*, II (1857-58), 69.

along a great arc that passed through northwestern California and down the line of the Sierras almost to the southern extremity of the San Joaquin Valley. This meant, he calculated, that mines were being worked over nearly nine degrees of latitude, or over a region seven hundred miles long by fifty miles wide—35,000 square miles.[10] It is obvious that 50,000 or even 100,000 men scattered over so great a surface area could hardly be regarded as a dense working population.

If their numbers were hardly adequate for the task at hand, the early gold seekers' technical knowledge was even less sufficient. The prevailing ignorance of geology and mineralogy was well illustrated by the homely attempts of Marshall, Sutter, and their employees to test the golden grains that they had found in the millrace.

Even after these experiments, Marshall's men did not appreciate the significance of their own discovery until they had shown specimens of the ore to Isaac Humphrey, who had at one time been a gold miner in Georgia. Humphrey, realizing at once that fortunes were in sight, hurried up to Coloma and inaugurated the first gold washing in California. He was soon joined by a Frenchman named Baptiste, who had been a gold miner in Mexico.[11]

It was fortunate for the development of far western

[10] William P. Blake, "Observations on the Extent of the Gold Region of California and Oregon," *American Journal of Science and Arts*, 2nd series, XX (1855), 73-74.
[11] Hittell, *Mining in the Pacific States*, pp. 14-15.

mining that there were such men as Humphrey and Baptiste. The average American, in 1848, was "handy with" a considerable number of trades and occupations, but mining was not one of them. Nor was it the business of most of the Europeans. By long established tradition, mining in the Old World was a profession which drew its ablest engineers and its most skillful workmen from certain famous localities that had contributed generation after generation of recruits. The duchy of Saxony, in southern Germany, the English county of Cornwall, and the Iberian Peninsula were perhaps the best known of these training grounds.[12] In these traditional centers, mining advanced to a remarkably high level during medieval and early modern times. The descriptions given in Agricola's great treatise of 1550 show clearly that by the early part of the Renaissance, Europeans had developed many of the fundamental techniques and tools that are in use today.[13]

The comparatively international mindedness of the miner seems to have been one of the important stimuli behind this progress. Apparently medieval and early modern miners were as willing as those of today to try their luck in a distant region if the opportunity seemed

[12] Thomas A. Rickard, *Man and Metals: A History of Mining in Relation to the Development of Civilization* (New York, 1932), II, *passim;* A. K. Hamilton Jenkin, *The Cornish Miner, An Account of his Life Above and Underground from Early Times* (London, 1927), *passim.*

[13] Georgius Agricola, *De Re Metallica, Translated from the First Latin Edition of 1556, with Biographical Introduction,* trans. by Herbert C. and Lou H. Hoover (London, 1912).

attractive. One hears repeatedly of Germans in England, and of skilled Saxons being imported into Hungary and Serbia. Agricola speaks of Italians coming to work in the German mountains, and Cornish legend insists that one of the fundamental machines used in vein mining was introduced into Germany by a Cornishman.[14]

By this migration of individuals and groups, the more progressive areas of Europe became the possessors not only of the lessons learned from their own experience, but also of the information gained by others in quite different parts of the civilized West. After the discovery of America the Spanish branch of this international fraternity served as the medium through which the accumulated knowledge of the Old World was transplanted to the New.

In America, Iberian and Ibero-American miners practiced their inherited trade in gold, silver, and mercury mines located in Peru, Chile, Bolivia, Colombia, Mexico, and Brazil. They had been operating at some of these mines, such as those at Potosí, for three hundred years by the time of Marshall's discovery. When word reached them of the great events at Coloma, they prepared to move northward, just as generation upon generation of their forebears had always been ready to respond to reports of a new "excitement."

This was not their first invasion of United States ter-

[14] Agricola, pp. 282-283 (fn. 8), 334; Jenkin, *Cornish Miner*, p. 92; Rickard, *Man and Metals*, II, 513, 521-523, 531, 539, 549.

ritory. At the close of the eighteenth century, gold was discovered in North Carolina, and subsequently in South Carolina, Georgia, Tennessee, and Alabama. There, an important gold mining industry developed during the half century that preceded Coloma. To it came many Americans, but also many Europeans and South Americans. Germans, Englishmen, Welshmen, Cornishmen, Italians, Hungarians, Mexicans, and Brazilians were among the races represented.[15]

Through the joint efforts of experienced Europeans and South Americans, and of inventive but untrained Americans, the basic problems of gold mining were successfully dealt with in the South during the twenties, thirties, and forties. When the news from California was confirmed at the close of 1848, the South was well prepared to join Latin America and Europe in providing the technical skill and implements needed in the new mineral area.

A few other sections of the United States also had enough familiarity with mining, in 1848, to make some contribution. Among the men joining in the rush to the Pacific Coast one hears of coal and iron miners from Pennsylvania, and lead miners from the upper Mississippi Valley.[16] The latter were especially important.

[15] Fletcher M. Green, "Georgia's Forgotten Industry: Gold Mining," *Georgia Historical Quarterly*, XIX (1935), 93-111, 210-228; same author, "Gold Mining: A Forgotten Industry of Ante-Bellum North Carolina," *North Carolina Historical Review*, XIV (1937), 1-19, 135-155.
[16] Henry De Groot, "Six Months in '49," *Overland Monthly*, 1st series, XIV (1875), 316.

Many of them were Cornishmen who had come to Wisconsin and other parts of the Northwest during the thirties and forties.[17]

With the arrival of European immigrants in the latter part of 1849, other Cornishmen, together with Germans, Spaniards, Welshmen, Frenchmen, and Italians, began coming directly to the new mecca from the most famous mining districts of Europe. California thus received, from both the Old World and the New, delegations of miners who possessed a considerable fund of experience and practical knowledge.

In the absence of both technical schools and reliable manuals,[18] these veterans became the teachers of the untrained Americans and Europeans who formed the bulk of the Gold Rush population. In forty-nine and the early fifties it was considered a great advantage to have a Georgian, Carolinian, or Cornishman in one's party, and more than one Anglo-Saxon acquired his first knowledge of gold mining by working alongside a Chilean or Mexican in the Sierra foothills.[19]

The surprising feature of the Gold Rush is not that

[17] Joseph Schafer, *The Wisconsin Lead Region,* Wisconsin Domesday Book, General Studies, III (Madison, Wisconsin, 1932), 44, 55-56, 106-108, 189, 213.

[18] In 1849 manuals suddenly became plentiful, but the quality of most was poor.

[19] *Sonora Herald,* November 9, 1850, June 5, 1852; Kelly, *Excursion,* II, 18-19, 23; Daniel B. Woods, *Sixteen Months at the Gold Diggings* (New York, 1851), pp. 123, 132; Stephen L. Fowler, "Journal of Stephen L. & James L. Fowler, of East Hampton, Long Island," p. 30 (typed copy of MS. journal, Bancroft Library, Berkeley, California).

HOW THE CALIFORNIA MINES ARE WORKED.

[Published at the WIDE WEST OFFICE, 151 Clay Street, San Francisco.]

farmers and clerks were so quickly metamorphosed into miners, but rather that it took so long for the full measure of Old and New World experience to be utilized. Machines and methods that had been tried successively in Europe, Latin America, and the southeast of the United States were one, two, and even three years in achieving widespread application in California. In some cases Americans of the Far West had to reinvent processes that, unknown to them, were in daily use in countries below the Rio Grande or across the Atlantic.

It is difficult to escape the conclusion that many lives, thousands of dollars, and several years of time could have been saved by a well-directed attempt to collect, codify, and publish the existing mining knowledge of Europe and America. Instead of doing so, men rushed to El Dorado and impatiently learned the simplest method by which they could begin mining with the least capital and the minimum of delay.

V

FROM AN ADVENTURE TO A BUSINESS:
MINING, 1848-1851

Fortunately for the Argonauts, all of the mining of 1848 and 1849, and most of that of the early fifties, was for placer gold rather than for the more difficult vein gold. For countless centuries the streams had been dropping flakes and chunks of the precious mineral onto sand bars, into crevices in the banks, and into "pot holes" in the beds of the rivers. Every stone along a river's course had provided an obstacle behind which fragments might lodge.

Water had been the primary agent in storing up this accumulation of wealth, and water was to be the miner's chief weapon in his attack upon it. Sometimes the miner might be fortunate enough to find crevices richly piled with comparatively pure gold dust that could be scraped out with a knife and spoon,[1] but in most cases he had to dig down into promising-looking ground with a pickaxe and shovel. After he had thrown up mounds of "dirt," he found himself facing the problem of separating the yellow grains from the gravel and sand with which they were intermixed.

Long generations of Europeans, Latin Americans, and

[1] Kelly, *Excursion*, II, 28, 106; San Francisco *Daily Alta California*, August 26, 1852.

Americans had worked out several different solutions to this problem. The simplest one involved the use of a flat-bottomed pan or a bowl with gently sloping sides. The southerners of the United States used the former, generally manufactured out of iron or tin; the Latin Americans the latter, fashioned out of wood. In either case the instrument was simply a modern version of the bowls described by Agricola in the sixteenth century.[2]

The operation of the pan was based upon one of the qualities of gold: its unusual weight. A given volume of gold far outweighs an equivalent volume of most kinds of rocks and earth.[3] It was this very peculiarity that had caused the streams to drop their precious freight instead of carrying it further in suspension. The miner's aim was to re-create and extend the natural process of sorting the heavier gold from the lighter debris.

The gold seeker would fill his pan or bowl with "dirt" and water. He would knead the two ingredients into a mixture, and pick out the larger stones. Then, grasping the rim of the vessel with his hands, he would thrust the pan under the surface of a pool or stream of water.

One side of the pan being held a little higher than the other, by a peculiar circular motion of the hands a revolving current is produced within it, which carries away the lighter portions

<hr />

[2] The Latin American bowl was called a *batea*. On the pan see "Mining for Gold in California," *Hutchings' Illustrated California Magazine*, II (1857-58), 4; Eugene B. Wilson, *Hydraulic and Placer Mining* (New York, 1901), pp. 21-25; D'Arcy Weatherbe, *Dredging for Gold in California* (San Francisco, 1907), pp. 16-17. Note Agricola, pp. 334-336.

[3] Lindgren, *Mineral Deposits*, pp. 221-222.

[of the "dirt"] over its top, while the heavier matters remain behind. In this way the earthy particles are gradually washed away, the pebbles being removed by hand, until nothing is left but the gold [which sinks to the bottom of the pan on account of its weight], either entirely clean, or mixed with a small quantity of heavy sand.[4]

This method had the advantage of simplicity. Anyone could learn it with a little patience. The chief complaint against it was its slowness and unsuitability to large-scale operations. The pan was a one-man instrument, and with it even the most skillful of miners could wash only a limited amount between sunrise and sunset.

As a step towards larger production, the miners turned to the "rocker" or "cradle," in the spring and early summer of 1848. This machine, like the pan, had long been in use in Georgia and North Carolina. Isaac Humphrey is said to have used both in his first operations at Coloma. The cradle much resembled the family institution from which it derived its name.[5] Like a child's cradle, it was an oblong box without a top and mounted on rockers; it was several feet in length and was placed

[4] Titus F. Cronise, *The Natural Wealth of California, Comprising Early History; Geography, Topography . . . Agriculture . . . Mineralogy, Mines and Mining* (San Francisco, 1868), p. 533.

[5] Descriptions of the cradle and the method of operating it are given in: "Mining for Gold," *Hutchings'*, II, 4-5; Col. R. B. Mason, Monterey, to Brig. Gen. R. Jones, August 17, 1848, "California and New Mexico, Message from the President," *House Exec. Doc.*, 31 Cong., 1 sess., no. 17 (January 24, 1850), p. 529; New York *California Herald-Extra*, January 16, 1849. On its antecedents in the South, see Charles E. Rothe, "Remarks on the Gold Mines of North Carolina," *American Journal of Science and Arts*, 1st series, XIII (1828), 208-209.

in a sloping position. At its foot it was left open, by the omission of the endboard; at its head a sieve or hopper was attached, and along the bottom of the cradle cleats were nailed at intervals.

To operate the machine, auriferous "dirt" was poured into the hopper, whose sievelike bottom kept back the larger stones while allowing the finer material to fall through to the bottom of the cradle. Meanwhile, water was bailed vigorously into the hopper, so that it would spill down with the "dirt" and create a current that would hurry the "dirt" along the bottom towards the open lower end. The gold, being more weighty than the rest of the debris, tended to drop behind and pile up in little drifts behind the cleats. In order to expedite the disintegration of the earthy material and the settling of the gold, the whole machine was kept swinging back and forth on its rockers, with each swing interrupted by a slight jerk.

The cradle was not overly efficient, for it permitted much of the finer gold to slide over the cleats and escape, but it did make possible gold washing upon a larger scale. Thereby it enabled groups of men advantageously to "club" together their efforts: one to rock the cradle and pour in water; one to haul the dirt and load the hopper; and one or more to shovel and dig.[6]

[6] Mason to Jones, August 17, 1848, "California and New Mexico," *House Exec. Doc.*, 31 Cong., 1 sess., no. 17, p. 529; New York *California Herald-Extra*, January 16, 1849; San Francisco *Daily Alta California*, August 26, 1852; Kelly, *Excursion*, II, 20.

In the pan and the rocker the gold seeker found the basic tools with which he was to work throughout the first two seasons of the Gold Rush.[7] At the same time, in the practiced skill of the Georgian, Carolinian, and Latin American he found competent models to imitate. Thus equipped, he began the exploitation of the virgin placer deposits that nature had been accumulating. Because he was the first, the miner of 1848 and the first six months of 1849 was able to confine himself to the easiest and most accessible of the diggings: the river bars and banks, the creeks and gulches, and those parts of watercourses that were annually laid bare by the long drought of the California summer.[8]

Here the miner found wealth in amounts so vast that by November 1848 the usually cautious military governor of the province was notifying his superiors that "any reports that may reach you of the vast quantities of gold in California can scarcely be too exaggerated for belief."[9] Individuals, in those prosperous days, made astonishingly lucky "strikes." A soldier was given twenty days' leave to go to the mines. He had to spend half of his furlough in traveling, yet during the remaining time he amassed $1,500—a sum greater than the

[7] J. Ross Browne and James W. Taylor, *Reports upon the Mineral Resources of the United States* (Washington, 1867), p. 16; hereinafter cited as Browne, *Report* (1867).

[8] "Mining for Gold," *Hutchings'*, II, 3; Hittell, *Mining in the Pacific States*, p. 21.

[9] Col. R. B. Mason, San Francisco, to Brig. Gen. R. Jones, November 24, 1848, "California and New Mexico," *House Exec. Doc.*, 31 Cong., 1 sess., no. 17, p. 648.

amount he would have received in pay, clothing, and rations during his entire enlistment. In a single day a soldier could earn in the mines more than double his pay and allowances for a month in the army.[10] At Coloma, men commonly made from $25 to $30 per day by working with the rocker, while the first to "work" the rich bars of the North Fork of the American and of the Yuba, Feather, Stanislaus, and Trinity rivers often secured $500 to $5,000 in a single day. It was not unique to average $300 to $500 per day for several weeks in 1848-49.[11]

Viewed as a whole, however, the 5,000 men who were engaged in mining at the close of 1848 probably did not have a higher daily average than $20, and in the years which followed this "wage" declined rather than increased.[12] In return for this sum, men subjected themselves to a grueling type of labor that they would have scorned in normal times. A forty-eighter said of it: "I found digging gold by no means the enchanting employment many might dream it to be; but a matter-of-fact, back-aching, wearisome work—most nearly resembling, for all the world, the heavy toil of a multitude of Paddies excavating a canal, or millrace." [13]

[10] Mason to Jones, August 17, 1848, *ibid.*, pp. 533-534.

[11] Hittell, *Mining in the Pacific States*, p. 21.

[12] *Ibid.*; cf. a contemporary statement that "gold digging at the present time [1848] yields a pretty sure income of $10 to $20 per day with the chance of making from $100 to $500 in the same time as is not infrequently done." Chester S. Lyman, "Conditions in California in 1848, A Letter to 'The Friend,' Honolulu, from Chester Smith Lyman," *California Historical Society Quarterly*, XIII (1934), 177.

[13] Lyman, *ibid.*, p. 177.

Gold mining meant swinging a pick, shovel, or crow-bar all day in the burning glare of the California sun, with the temperature up above one hundred degrees. It meant digging through gravel and sand until one's back ached and his shoulders cried out in rebellion. It meant slogging about in cold, muddy water until one's feet were numb and his shoes were a soggy, useless wreck. Or one might have to spend hours kneeling or stooping over while he scraped out crevices, or anxiously washed the "dirt" in his pan. Still others had to watch the day-light hours pass while they monotonously rocked the cradle to and fro and poured in "dirt" and water.[14]

Some were too restless to endure such a life. They felt sure that if they ventured beyond the boundaries of the known auriferous area they could find new, untouched "diggings," where nuggets could be had for the picking up. So they organized "prospecting" expeditions. They would load a pick, shovel, pan, some food, and a gun onto a mule or onto their own backs, and start off across difficult ravines, gulches, and mountains, and in the face of Indian danger, in search of virgin ground.[15]

Such trips rarely produced a fortune for their or-ganizers. Too much time was lost in wandering about and taking countless "prospects" with the pan at every

[14] Mason to Jones, August 17, 1848, "California and New Mexico," *House Exec. Doc.*, 31 Cong., 1 sess., no. 17, pp. 529-530; New York *California Herald-Extra*, January 16, 1849; San Francisco *Daily Alta California*, August 26, 1852; cf. p. 86 below.
[15] "Mining for Gold," *Hutchings'*, II, 3.

promising-looking stream. Yet always the prospector was buoyed up by the conviction that *if* he should stumble upon another American, Feather, or Yuba River, he, too, would earn $500 or $5,000 in a single day.

Beginning in 1848 and 1849, and continuing throughout the fifties, these restless optimists roamed over the mountainous face of Sierran and northwestern California. Individually they earned far less than those who were willing to submit to the back-breaking routine of the known diggings, but collectively they enriched the whole world, for by their discoveries they revealed to others the location of widely scattered auriferous streams and gulches.[16]

The confirmed prospectors were not the only restless members of the early California mining community. As the great influx grew in volume during 1849, the veterans of the previous year began to feel crowded. Like the frontiersmen who could not endure the presence of a neighbor in their wilderness, many took refuge in flight. They abandoned the known diggings to the newcomers and themselves moved on in search of virgin ground.[17]

Their reason for so doing was that in the older districts the average yield per man was less in 1849 than it

[16] Doings *(pseud.)*, "Prospecting," *Hutchings' Illustrated California Magazine*, I (1856-57), 554-555; "The Prospector," North San Juan *Hydraulic Press*, December 4, 1858.

[17] San Francisco *Alta California* (steamer ed.), August 2, 1849.

had been in 1848, because there was a larger number of persons to claim the same extent of auriferous ground.[18] The more progressive of the forty-eighters and forty-niners, however, met this condition of increasing population and declining yield by a better method than retreat: by improving the methods of mining.

One obviously needed improvement was in the final extraction of gold after the "dirt" had been washed. With either the cradle or pan, the final residuum after washing was not pure gold. Usually there was a considerable admixture of heavy black sand which weighed too much to be separated from the gold by the current of water. The Latin Americans among the early Argonauts had long been familiar with this problem, and they contributed the first method for getting rid of the sand. It was their custom to dry the mixed gold and sand in the sunlight, or over the fire, then to separate the two by blowing upon the dry grains or by tossing them up into the air as one would wheat and chaff—trusting to the weight of the gold to keep it from being winnowed away with the sand.[19]

The waste caused by this reliance upon one's lung power or the wind was obvious. A much better way to separate the gold from the sand was to use mercury (quicksilver). This was an ancient technique, known

[18] See table of average yield per man, below, p. 120.
[19] Moerenhout, *Inside Story*, pp. 17-18; Fowler, "Journal," p. 40; Bayard Taylor, *Eldorado; or, Adventures in the Path of Empire . . . Pictures of the Gold Region* (Household Ed.; New York, 1882), p. 89.

not only to Agricola but even to Pliny. It was long employed in Latin American mines and was used in North Carolina as early as 1809.[20]

As the medieval alchemists would have explained it, mercury has an "affinity" for gold. By placing the two in contact, the mercury will incorporate the gold particles into itself while excluding foreign matter such as sand. The resultant amalgam can then be reduced to its two constituents by being heated in an enclosed vessel, for the mercury will vaporize and leave behind it the gold, free of impurities.

The miners began making use of this fortunate chemical relationship in the latter part of 1849. They found that the mercury could be employed after the residuum of sand and gold had been obtained, or, more satisfactorily, that it could be placed in the bottom of the rocker, where it would exercise its "affinity" upon the gold when the latter slid into small drifts behind the cleats.[21]

At about the same time men began to experiment with an exploitative method that can hardly be called an invention, but which was certainly one of the most striking developments achieved by California mining: the

[20] Agricola, pp. 295-300; Pliny, *The Natural History of Pliny*, trans. by John Bostock and Henry T. Riley (Bohn's Classical Library; London, 1855-1857), bk. xxxiii, chap. 32; Green, "Gold Mining," *North Carolina Historical Review*, XIV, 8.

[21] On first use in California, see Hittell, *Mining in the Pacific States*, pp. 22, 133; Taylor, *Eldorado*, p. 258; San Francisco *Weekly Pacific News* (steamer ed.), March 1, 1850. On method of using it, see Wilson, *Hydraulic and Placer Mining*, pp. 27, 32-34.

diversion of rivers from their natural beds so that they might be robbed of the gold that had accumulated in "pot holes" and cracks beneath their waters. The miners of 1848 had been content with the wealth that lay along the streams' edges and on exposed bars, but as early as July 1849 at least two associations of men were engaged in damming parts of the American River, preparatory to turning the water into artificial channels.[22]

Such attempts at hydraulic engineering automatically necessitated the union of many men's efforts towards a common end. Hired labor was hardly to be had at all in the mines at first, and what was available was so costly as to prohibit its extensive use. Men therefore resorted to joint stock associations in which the "subscriptions" were paid in the form of labor. The members of each association bound themselves to work upon their project every day, under the direction of their elected officers and under penalty of fines for unauthorized absence from the scene of operations. The proceeds were to be divided equally among the members.[23]

This provided a method for "getting at" an entirely new source of gold, just as the use of mercury was making possible the saving of a larger percentage of the gold contained in any "dirt" that was washed. There re-

[22] San Francisco *Alta California* (steamer ed.), August 2, 1849.
[23] "Copy of the Constitution of the North fork Dam & Mining association, June 27, 1849," Papers of James L. L. Warren, 1849-1890, box 5 (Bancroft Library, Berkeley, California); "Articles of Agreement of the Hart's Bar Draining and Mining Company," July, 1850, text in Woods, *Sixteen Months*, pp. 144-148.

mained a third possible improvement: the attainment of something approaching mass production in the handling of auriferous material, so as to reduce the cost per cubic yard and thereby render profitable the working of the less attractive deposits.

In answer to this need the "long tom" was introduced during the winter of 1849-50. This instrument had been used for twenty years in Georgia and North Carolina, and one of the machines described by Agricola sounds singularly like it.[24] It was an outgrowth of the cradle. It was built in two sections: a top and a bottom. The top section was an open-ended box or trough, usually about twelve feet long and always mounted in a down-sloping position. This trough was shaped like an inverse funnel: starting with a width of one or two feet at the higher end, it increased to double that width at the middle. At the middle the sides straightened out and ran parallel until they reached the lower end. In this broad, straight part the bottom was made of perforated sheet iron, with the sheet iron bent up into a gradual curve as it approached the lower end.

The bottom section of the long tom was placed underneath the perforated sheet iron. It was called a riffle box,

[24] For first use in California, see *Sacramento Transcript*, April 1, 1850; *Sacramento Daily Union*, January 15, 1855, correcting *Golden Era;* Browne, *Report* (1867), pp. 18-19. For earlier use elsewhere, see Green, "Georgia's Forgotten Industry," *Georgia Historical Quarterly*, XIX, 217, and his "Gold Mining," p. 12; also Agricola, p. 323. For description of it, see "Mining for Gold," *Hutchings'*, II, 5-6. Note that the long tom was stationary, in contrast to the cradle.

and was simply a wooden box with cleats ("riffle" or "ripple" bars) across its bottom. To operate the machine water was piped into the high, narrow end of the trough. As this stream cascaded down the inverse funnel, two men continuously shoveled dirt into it, while a third man stood by the perforated sheet iron portion to stir the mixture as it piled against the up-curving strainer. In this way the coarser parts of the debris were retained in the top section of the long tom, while the finer material plunged through into the riffle box, where the gold was separated from the rest of the debris in precisely the same fashion as in the bottom of the cradle.

Obviously the long tom was simply an adaptation of the cradle to the needs of large-scale production. The top section, or trough, of the long tom was the hopper of the cradle enlarged so as to permit the use of a continuous current of water instead of bailing by hand. So much larger a volume of dirt could be handled that two or more men could be uninterruptedly employed in shoveling and a third in stirring. The bottom section, or riffle box, was essentially the cradle with the hopper removed.

This severance of the riffle box from the hopper led to a further improvement a year later, in the winter of 1850-51. Realizing that the long tom let many of the finer particles of gold escape, the miners tried placing a series of riffle boxes, in a chain, at the outflow end of the long tom. By so doing they subjected the escaping fine debris to several successive washings. When this experiment brought satisfactory results, the "sluice" had come into being. The sluice, or sluice box, was simply an open trough, twelve or fourteen feet long, in the bottom of which were placed either riffle bars or else false bottoms so perforated and split as to provide crevices in which the gold particles would lodge. Usually a number of sluice boxes were arranged in a "string," with the lower end of one fitted into the upper end of the next one. As with the cradle and long tom, mercury was customarily used in the sluice to help catch and separate light flakes of gold that would otherwise be swept over the riffles and lost.

The historical origin of the sluice is a matter of some dispute. Unquestionably the device has an ancient history that spans at least two continents, but there is reason to believe that it was reinvented in California by men who had no knowledge of its earlier use elsewhere.[25]

Sluices were used both in conjunction with the long tom and as an independent instrument into which dirt was shoveled directly. In the latter case, especially, the sluice became a further step toward the handling of

[25] On its California origin, see Hittell, *Mining in the Pacific States*, p. 22, and Browne, *Report* (1867), p. 19. On its antecedents, see Pliny, *Natural History*, bk. xxxiii, chap. 21; Agricola, pp. 321-336; Green, "Georgia's Forgotten Industry," p. 216. For description of it, see "Mining for Gold," *Hutchings'*, II, 6-8; Cronise, *Natural Wealth*, pp. 533-534. The use of mercury was essential when the dirt contained fine gold that would otherwise be swept away by the rush of water. Note that experiments with sluices were made as early as the winter of 1849-50.

larger masses of auriferous material. Only one man at a time could use the pan. A single man could operate the rocker, but he could associate with him the labor of one, two, or even three others. With the long tom three to six men could be kept steadily at work, while with the sluice from five to twenty men could combine their efforts profitably.[26]

Both the long tom and sluice required a continuous stream of water. If a group of miners were lucky, the ground they were working would be located so near to a stream that water could be piped to the "tom" or sluice through a few feet of ditching, wooden fluming, or canvas hose. Frequently no such easy solution was possible, and in 1850 men began the construction of systems of ditches and flumes designed to convey water across the country to areas that had been left dry by nature.

The earliest of these was undertaken in Nevada County in 1850. It brought water to the Coyote Hill diggings from a creek one and a half miles away. Like the river damming enterprises, the ditch and flume projects were generally undertaken by joint stock companies, but the working membership in them seems to have been smaller than in the damming associations. In a number of instances they were originated by individuals or partnerships. Frequently the local merchants in the mining towns subscribed to the stock of the companies, and with the funds which they provided hired labor was

[26] Hittell, *Mining in the Pacific States*, pp. 132-136.

secured, even at the almost prohibitive rate of eight dollars per day for common labor. The profit from the ditch companies came through selling to the miners the use of the water thus obtained.[27]

If regarded individually, the use of mercury, the introduction of the long tom and sluice, and the beginning of damming and ditching were each of great importance. Collectively, however, they constituted a complete revolution in mining. They represented improvements at almost every stage in the process of reducing natural deposits to marketable gold dust. The employment of mercury improved the efficiency of all washing operations. The long tom and sluice made possible the handling of larger masses of dirt at a lower cost per cubic yard. The canals and ditches supplied the water that was needed for the tom and sluice. By reducing the cost per unit of material handled, the several improvements permitted the miners to extend their work into comparatively low-grade auriferous ground that had not previously been considered rich enough, while the damming and diversion of rivers laid bare an entirely new source of gold.

[27] T. H. Rolfe, "Mines and Mining," in *Bean's History and Directory of Nevada County, California, Containing a Complete History . . . with Sketches of the Various Towns and Mining Camps,* comp. by Edwin F. Bean (Nevada City, California, 1867), pp. 65-72; John S. Hittell, *The Resources of California, Comprising the Society, Climate, Salubrity, Scenery, Commerce and Industry of the State* (6th ed.; San Francisco, 1874), pp. 305-306; San Francisco *Daily Alta California,* December 4, 1852.

The result of this technical progress was that the miner found that his prosperity did not cease with the inevitable fading of the rosy flush that had marked the dawn of the Gold Rush. According to California's leading newspaper, by midsummer of 1851 the easily worked placers were losing "much of their attractiveness, even to new-comers to the land of gold," because of their "positive decline" in yield.[28] Indeed, the *Alta California* was almost ready to make the premature confession that "the excitement relative to California gold is now over." [29] The truth was, however, as the *Alta California* pointed out on another occasion, that mining was still successful, and for this reason:

The secret of the success is mainly to be found in the added science and skill which have been brought to bear upon the auriferous region through the experience of former operations. . . . the miners are beginning to discover that they are engaged in a science and a profession, and not in a mere adventure; and they have accordingly set themselves to the task of acquiring that knowledge of their art which is the first condition of success in all undertakings.[30]

In other words, mining was coming of age.

[28] San Francisco *Daily Alta California*, July 17, 1851.
[29] *Ibid.*, September 15, 1851.
[30] *Ibid.*, September 23, 1851. It should be remembered that there was inevitably a time lag in the spread of mining improvements. J. Ross Browne said that "in most districts" the long tom and sluice "were unheard of until late in 1850 or 1851." Browne, *Report* (1867), p. 19. Thus the full effect of the new methods would not have been felt until the mining season of 1851.

LIFE AMONG THE MINERS.

Seen here are many changing scenes
    Met with in a miner's life,
Some of his comforts and his joys,
    Some of his toils and strife;
His life is one of hard, unceasing toil,
    A truthful tale is told
Of joys and sorrows, incident
    To those who dig for gold.

His cabin built of logs, and in
    A quaint, primeval style,
Intended but to shelter him,
    Until he makes his pile,
We see the miner hard at work,
    As steady as a saint—
His ground is rich, and he has got
    Poor ground to make complaint.

This washing dirt for gold is well,
    When well they make it pay;
But few attractions unto them
    Is the red shirt washing day.
Upon a bed of sickness, now,
    No loving friend is there;
How much he needs a sister's aid,
    A mother's anxious care.

Saturday night they weigh their dust,
    All anxious faces there;
While waiting for the truthful scales,
    To give to each his share.
Letters from *home*—there's nought can give
    The miner joy like this—
Good news from loved ones, far away,
    Is life, extasy, and bliss.

—from *Put's Original California Songster*, originally published in several successive editions in the 1850's. "Old Put" was John A. Stone. His career, like that of so many who took part in the Gold Rush, included moments of success but ended in final failure. According to a reminiscent account published in 1886, Stone was a singer and song writer who was a general favorite throughout the mines. With a troupe called the Sierra Nevada Rangers he gave concerts in the principal mining towns, while his songs, published in booklet form, sold by the thousand. Through these two sources of income he could easily have achieved a modest competence and returned to his wife and family in the East, but liquor and gambling took his money almost as quickly as he earned it. In between concert tours he lived in a cabin in Greenwood, El Dorado County. On occasion he worked as a miner. He prided himself on knowing from personal experience the hardships and disappointments met with in the life of the average miner.

VI

CAMPS, CABINS, CITIES, AND CITIZENS
1848-1858

For whatever progress mining achieved during its early years, the credit must go to that sturdy individual upon whose efforts the whole industry depended: the average miner. The latter might be a person of almost any national origin. He might be an American, a European, a Latin American, or an Asiatic. In the eyes of contemporaries, however, he had to be an American or European to be regarded as typical, since there was always a pronounced tendency to consider as special groups the Chinese and Latin Americans.

Within the ranks of the Americans and Europeans a surprising degree of homogeneity was said to exist, thanks to the common sharing of common problems under conditions of extreme difficulty.[1] It was said also that, regardless of his background, after a few months in the mines each miner came to look and dress very much like all the rest. The universal costume was a red or gray flannel shirt, old trousers, high boots that were pulled up over the pant-legs, and a dilapidated slouch hat.[2]

[1] *Sacramento Transcript*, August 24, 1850; "The Mountaineers of California," *Hutchings' Illustrated California Magazine*, IV (1859-60), 76-77.
[2] John H. Eagle, Auburn Ravine, California, to his wife, June 28, 1852 (Correspondence of John H. Eagle, in the Huntington Library, San

Because they were so far from feminine civilization, and because they were living a life that was hostile to neatness and foppery, most of the miners allowed their appearance to deteriorate into picturesque disreputableness. A stranger would have concluded that "the miners of California never shave; never put on clean vests, clean dickeys, or clean boots." [3] He would doubtless have described them as "a gang of Irish-appearing, hard-working miners, habited in their red flannel shirts, rough as the grisly [sic] bear, long beards, long hair, old hats, no shoes, or shoes variously patched." [4]

Their places of habitation were as casual as their dress. When the Gold Rush began, California had little community life around which to build. At San Francisco there was a nascent seaport colony. On the site of the future cities of Sacramento and Stockton there were a few buildings. Near San Francisco Bay there were the farming hamlets of San José and Sonoma, and on Monterey Bay there were the provincial capital, Monterey, and an agricultural settlement, Santa Cruz. Elsewhere there were only isolated ranches.

When thousands of people suddenly came hurrying into this wilderness, settlements began to spring up at

---

Marino, California; hereinafter cited as Eagle Correspondence); North San Juan *Hydraulic Press*, December 4, 1858; "The Mountaineers," *Hutchings'*, IV, 76.

[3] *Ibid.*

[4] Woods, *Sixteen Months*, p. 125. For the two extremes, the dandy and the gaunt "Pike" backwoodsman, see Borthwick, *Three Years*, pp. 147-148.

all kinds of places, apparently in response to no guidance save that of chaos. Gradually, however, the many new communities began to assume a logical relationship towards one another. It became apparent that San Francisco was to be the queen city of El Dorado. Upon its wharves must be landed all of the goods and most of the travelers that came to California.

Sacramento and Stockton were to play a different role. They were river ports, and as such were subordinate in importance to San Francisco. They were, respectively, the chief cities on the Sacramento and San Joaquin rivers. Ranking with them was Marysville, a new community that was established at the close of 1849 at the head of navigation on the Feather River, the main tributary of the Sacramento.

Save for the overland caravans from Oregon, Mexico, and the Mississippi Valley, supplies and immigrants from the East and Europe usually were landed at San Francisco from ocean-going vessels, and there were transferred to river craft—light-draft steamboats and small sailing vessels—for reshipment to the three interior commercial cities of Sacramento, Stockton, and Marysville. From those three points both goods and gold hunters had to travel overland for another fifteen to fifty miles before reaching their destination. If a miner were affluent, he could make this last part of the journey in a crowded stagecoach that would clatter over the rough roads at an excitingly dangerous speed; if he were "down on his luck," he would have to walk. His supplies would come to him by slow mule or ox teams, unless he decided to make his headquarters far up in the remote and mountainous diggings whither only pack trains could go. His mail and newspaper would come to him more rapidly, thanks to the efficient express companies, and his gold, when he had been so fortunate as to find it, would be sent back to San Francisco in the custody of the same organizations.

In the mineral region itself there were several types of communities that served the miner. The most common was the "camp": a straggling settlement that might vary in size from a few houses to a small town. A more impressive place was the mining town, a community that was larger in size than the camp, had a somewhat less ephemeral life, and usually had a few buildings that could make some pretensions to substantiality. A few of the mining towns had an especial importance because they stood astride local trade routes. Such towns grew into small cities. Placerville, Nevada City, and Sonora were examples.

The average miner lived at or near a camp or mining town. The structure that he called home was usually a very crude affair. During the first two seasons, 1848 and 1849, nearly everyone was camping out, trusting to California's reliable climate for protection from sudden changes of weather. The only shelter was that provided by an overhanging tree, or a homemade tent, or a brush lean-to. As the winter of 1849-50 approached, the wiser Argonauts began to build log cabins, in anticipation of the rainy season. By the spring of 1850 all manner of structures might be seen:

The houses are of every possible variety, according to the taste or means of the miner. Most of these, even in winter, are tents. Some throw up logs a few feet high, filling up with clay between the logs. The tent is then stretched above, forming a roof. When a large company are to be accommodated with room, or a trading depôt is to be erected, a large frame is made, and canvas is spread over this. Those who have more regard to their own comfort or health, erect log or stone houses, covering them with thatch or shingles. . . . Some comfortable wigwams are made of pine boughs thrown up in a conical form, and are quite dry. Many only spread a piece of canvas, or a blanket, over some stakes above them, while not a few make holes in the ground, where they burrow like foxes. . . . The Mexicans and Chilinos put up rude frames, which they cover with hides.[5]

In the seaport and river cities and the larger mining towns, buildings of a more substantial type soon began to appear. During 1849, San Francisco, Sacramento, and Stockton were a jumble of canvas houses, beached ships, and ready-made imported wooden and metal buildings. Great "changes and improvements" appeared in San Francisco during the latter half of 1849, with "tents and canvas houses" giving "place to large and handsome edifices, . . . new hotels opened, market houses in opera-

[5] Woods, *Sixteen Months*, p. 121, entry dated April 2, 1850. Few spent the winter of 1848-49 in the mines.

tion and all the characteristics of a great commercial city fairly established." [6] Sacramento had a similarly lusty growth during 1849 and the following year. On the first of April 1850 its little newspaper declared:

We are struck with the contrast which a few weeks have produced. The din, the bustle, the confusion of business is witnessed in every quarter. —Ships and steamers, filled with passengers, are daily reaching our port, crowds of new comers throng the streets, and the entire city is a scene of life and activity. Meantime, in every direction, substantial buildings are springing up with wonderful rapidity,—tents are being removed, and we observe fine brick buildings erected and in progress of erection. A commodious theater, too, has lately thrown open its doors, and another will soon follow its example. Public edifices are being completed, and up and down the river, and broadcast far back upon the plains, the young city of Sacramento is spreading with a rapidity unexampled and unprecedented.[7]

Marysville, being the youngest of the commercial cities, was still "a little town, built mostly of canvas," in 1851. Three years later it, too, was "a large city," with "blocks of brick, fire-proof buildings." In 1851 Marysville had but two brick buildings; in the next year, seventeen; in 1853, thirty-nine; in 1854, forty-three; and in 1855, fifty-four.[8]

[6] Taylor, *Eldorado*, pp. 203-204, describing his visit in October, 1849.

[7] *Sacramento Transcript*, April 1, 1850; and cf. *ibid.*, November 16, 1850, with Kelly, *Excursion*, II, 69, 228.

[8] Quotations from Dolly B. Bates, *Incidents on Land and Water, or Four Years on the Pacific Coast* (5th ed.; Boston, 1858), p. 272. Statistics from William H. Chamberlain and Harry L. Wells, *History of Yuba County, California* (Oakland, California, 1879), p. 46.

Unfortunately for the safety of the populace, construction costs were so high and men's impatience so great that until late in the fifties the majority of the houses, whether in city, town, or camp, continued to be built chiefly of wood and canvas. In the cities and mining towns a favorite type of structure was a light wooden frame house in which tightly stretched cotton cloth was made to serve for interior partitions, ceilings, and wall coverings.[9] At the innumerable small camps men continued for several years to live in "round tents, square tents, plank hovels, log cabins, &c.,"[10] but in the middle and later fifties the standard habitation came to be a cabin with log or board sides, a canvas roof, and a mud-and-stones chimney.[11]

Such methods of construction produced communities that were veritable tinder boxes. Terrible fires wiped out most of the cities, towns, and camps several times over during the first decade of their existence. Since insurance facilities were not available until the later fifties, a conflagration usually inflicted a loss of 100 per cent upon those whose dwellings and stores were caught

[9] Bates, *Incidents on Land and Water*, pp. 184-185.
[10] Louise A. K. S. Clappe (Dame Shirley, *pseud.*), *The Letters of Dame Shirley: California in 1851-1852*, ed. by Carl I. Wheat, Rare Americana Series, ed. by Douglas S. Watson, nos. 5-6 (San Francisco, 1933), I, 37-38, dated Rich Bar, September 20, 1851.
[11] "Cabin Homes," *Hutchings' Illustrated California Magazine*, III (1858-59), 343-345; "An Evening Scene," *ibid.*, II (1857-58), 203; Prentice Mulford, *Prentice Mulford's Story: Life by Land and Sea* (The White Cross Library; New York, 1889), p. 116; John H. Eagle, Auburn Ravine, California, to his wife, November 14, 1852, Eagle Correspondence.

WHEN I WENT OFF TO PROSPECT.

I heard of gold at Sutter's Mill,
At Michigan Bluff and Iowa Hill,
But never thought it was rich until
    I started off to prospect.
At Yankee Jim's I bought a purse,
Inquired for Iowa Hill, of course,
And traveled on, but what was worse,
    Fetched up in Shirt-tail Cañon.

[CHORUS:]

A sicker miner every way
Had not been seen for many a day;
The devil it always was to pay,
When I went off to prospect.

When I got there, the mining ground
Was staked and claimed for miles around,
And not a bed was to be found,
    When I went off to prospect.
The town was crowded full of folks,
Which made me think 'twas not a hoax;
At my expense they cracked their jokes,
    When I was nearly starving.

[CHORUS:]

I left my jackass on the road,
Because he wouldn't carry the load;

76

I'd sooner pack a big horn toad,
      When I went off to prospect.
My fancy shirt, with collar so nice,
I found was covered with body-lice;
I used unguentum once or twice,
      But could not kill the grey-backs.

[CHORUS:]

At Deadwood I got on a tight—
At Groundhog Glory I had a fight;
They drove me away from Hell's Delight,
      When I went off to prospect.
From Bogus-Thunder I ran away—
At Devil's Basin I wouldn't stay;
My lousy shirt crawled off one day,
      Which left me nearly naked.

[CHORUS:]

Now all I got for running about,
Was two black eyes, and bloody snout;
And that's the way it did turn out,
      When I went off to prospect.
And now I'm loafing around dead broke,
My pistol and tools are all in soak,
And whisky bills at me they poke—
      But I'll make it right in the morning.

[CHORUS:]

—from *Put's Original California Songster*

in the path of the fire. Nevertheless, if a community had good prospects as a mining or commercial center, reconstruction would begin before the embers were cool.[12]

Socially the cities played a somewhat different role from that of the mining towns and camps. To the cities the weary miner came for his occasional spree, especially during the season of inclement weather, and for the miner's benefit the cities provided all manner of attractions. They had stores that carried imported necessities and luxuries. They had open-air auctioneering stands and horse markets, at which the gullible could be badly "stung." Hotels, often with very inferior accommodations, and restaurants of several nationalities catered to the visiting miner, while glittering saloons and gaudy "hells" helped him to spend his hard-earned gold.[13]

Such urban visits were the exception in the life of the average miner. For him, during the greater part of the year, the center of civilization was the nearest town or camp. It was there that he purchased his supplies and there that he found what little social life was available to him.

To reach "civilization" the miner might have to

[12] *Sacramento Weekly Union*, November 20, 1852, July 12, 1856; North San Juan *Hydraulic Press*, February 18, 1860; John F. Carrere, "Insurance in California," *Pacific Underwriter and Banker*, XXXIX (1925), 290-293.
[13] Bates, *Incidents on Land and Water*, pp. 118-122, 137-138; Hinton Rowan Helper, *The Land of Gold, Reality versus Fiction* (Baltimore, 1855), pp. 45-85, 131-146; John S. Hittell, *A History of the City of San Francisco and Incidentally of the State of California* (San Francisco, 1878), pp. 206-207, 214-215, 235-237; Borthwick, *Three Years*, pp. 47-73; Taylor, *Eldorado*, pp. 223-226.

struggle across several miles of rugged countryside. The site for a camp was usually determined primarily by proximity to gold diggings, so that ease of access received little consideration. The earliest camps were built amidst a topography that was hilly rather than mountainous, but many of the subsequent ones were located at the bottom of steep ravines and gulches. At some, the steepness of the surrounding ridges was so great that the rays of the sun never touched the community during the winter months.[14] At others, access to the camp was possible only by roads so abrupt that the approaching traveler found himself beginning the last stage of his journey from a point almost directly above his destination.[15]

Little effort was made to achieve beauty, comfort, or convenience in the arrangement of the camps. The camps were picturesque only in the extreme carelessness that characterized them. Buildings were strewn about in a most haphazard fashion, and ugly and dangerous mining excavations were left gaping vacantly between the houses.[16]

In each camp the single important street was usually

[14] Borthwick, *Three Years*, pp. 259-260; Clappe, *Letters of Dame Shirley*, I, 72, dated Indian Bar, October 7, 1851.
[15] Pringle Shaw, *Ramblings in California, Containing a Description of the Country, Life at the Mines, State of Society* (Toronto, c. 1858-1860), p. 58; and cf. "Mountain Scenery in California," *Hutchings' Illustrated California Magazine*, I (1856-57), 193-200.
[16] Clappe, *Letters of Dame Shirley*, I, 71, dated Indian Bar, October 7, 1851.

deeply rutted, and thick with dust in summer and with mud in winter. Because of the rugged topography, it was often a narrow, twisting, tortuous affair. Because of the habits of the men who used it, it was apt to be littered with old bottles, tin cans, and castoff clothing. The most prominent building that fronted upon it was nearly always a drinking and gambling resort that was often a hotel, and sometimes a brothel as well. Such an institution might be a bleak and unpainted eyesore on the outside, but inside it usually sported long mirrors, flashing stacks of glassware, and bright red calico curtains and trimmings.[17]

The big day in such a community was Sunday. That was when all of the miners who lived in the camp and the many others who lived within walking distance of it ceased work and gathered in town. Their ostensible purpose was to sell their gold dust and purchase supplies and equipment. The latter they could obtain from the many Yankee and German storekeepers and from the ubiquitous Jewish clothing merchants.[18]

The marketing, however, was only one of the reasons for the Sunday pilgrimage. After a week of hard labor,

[17] *Ibid.*, I, 27-30, dated Rich Bar, September 15, 1851, I, 37, dated Rich Bar, September 20, 1851, I, 73, dated Indian Bar, October 7, 1851; Borthwick, *Three Years*, pp. 112-114; Shaw, *Ramblings in California*, pp. 58-59.
[18] Borthwick, *Three Years*, pp. 116-119; Frank Marryat, *Mountains and Molehills or Recollections of a Burnt Journal* (London, 1855), pp. 262, 272; Henry Cohn, "Saint Louis and Poker Flat in the Fifties and Sixties. From the Jugenderinnerungen of Henry Cohn," trans. by Fritz L. Cohn, *California Historical Society Quarterly*, XIX (1940), 292-293.

PLACERVILLE

often in an isolated part of the mines, the visit provided a welcome opportunity for sociability, entertainment, and excitement. The miner was sure to find acquaintances among the crowd that thronged the narrow streets. He could renew old friendships at the numerous bars. Before the day was done he and his fellows would probably try their luck at the gaming tables, and there would lose a large part of the money they had earned so laboriously. The pool rooms and bowling alleys were certain to be well patronized. Usually there was a good audience for any itinerant theatrical performance, minstrel show, or circus that was passing through town. Sometimes there were horse races, on which everyone could make his bet, and occasionally there was a gory bull-and-bear fight—a Mexican contribution. At some towns there was a church service to which the religiously inclined minority might go.[19]

The carnival nature of the California sabbath was a clear indication of the abnormality of social conditions during the first decade after the gold discovery. As one distressed clergyman expressed it, when men stepped ashore at San Francisco, they entered a new world in which life was governed by new impulses;[20] they entered a world in which the consuming interest of all

[19] Borthwick, *Three Years*, pp. 118-120, 290-300, 334-340; Marryat, *Mountains and Molehills*, pp. 262-264, 272; Mulford, *Prentice Mulford's Story*, pp. 122-128; Cohn, "Saint Louis and Poker Flat," pp. 292-293; *Sacramento Weekly Union*, May 22, 1852; *Grass Valley Telegraph*, October 27, 1853.
[20] Woods, *Sixteen Month*, p. 47.

was the pursuit of sudden fortune, and in which there was a general absence of that sense of social position which guides men in normal communities.

The motley assemblage had come together from the ends of the earth. They were strangers to one another, and thus it mattered little what opinion one held of another. Everyone was busy with his own concerns; no one had time to worry about his neighbor's failings. Few had any desire to carve out for themselves a permanent career in California. The goal of all was a quick fortune and a speedy return "home"—to the East or to Europe or Latin America or Asia.[21]

The softening and restraining influence of family life was almost entirely wanting. According to the census of 1850, nearly 92½ per cent of the recorded population was male.[22] As the years passed the proportion became less unbalanced, but even in 1859 a well-informed writer estimated that in the mineral region there were six men to every woman, and he pointed out that of the women a considerable percentage were "neither maids, wives, nor widows."[23]

This necessarily implied that there were very few

[21] John S. Hittell, "The Necessity of Selling the Mineral Lands of California," *Hesperian*, III (1859-60), 495; Borthwick, *Three Years*, p. 67; Kelly, *Excursion*, II, 260.

[22] *Seventh Census, 1850*, p. 972.

[23] Hittell, "Mineral Lands," pp. 493-494. Another writer said the average was one to seven; *Sacramento Weekly Union*, February 12, 1859. Still another said one to ten; "Our Social Chair," *Hutchings' Illustrated California Magazine*, V (1860-61), 73.

family homes, and that fact, in turn, meant that except for his own rude and lonely cabin there was no place to which a man could go save to a gambling house or saloon. The latter, as one miner explained to his absent wife, were the real "public houses" of California. They were the first places to which a stranger would turn when searching for a friend. They were the only places where the tired miner could be sure of finding music, diversion, and the congeniality that arises from the presence of a crowd.[24]

Under such conditions vice was made easy, and the weaker of the Argonauts, "on arriving here . . . laid aside every rule which before had governed them, and, patterning after the Indians or the Mexicans, transformed themselves into squalid wretches, and too frequently gave an unbridled rein to all their passions."[25] The result of dissipation might be seen in the unpleasantly large number of men with broken-down constitutions and ambitions.[26] It might be seen also in the speed with which old age overtook men in California, although in that respect the sheer laboriousness of life in the mines was certainly an equally important factor. When the Gold Rush began, most of the Argonauts were "young

[24] John H. Eagle, Gold Hill, California, to his wife, December 25, 1852, Eagle Correspondence. Cf. Mulford, *Prentice Mulford's Story*, pp. 124-125, and San Francisco *Daily Alta California*, December 23, 1852.

[25] San Francisco *Daily Alta California*, December 12, 1852.

[26] Shaw, *Ramblings in California*, pp. 58-62; San Francisco *Daily Alta California*, July 27, 1855; *Napa County Reporter*, July 2, 1859, November 12, 1859; Columbia *Tuolumne Courier*, October 10, 1863.

men in the prime and vigor of life," [27] but as hardship and dissipation pressed harshly upon them, they aged more rapidly then the mere passing of the years. "Nowhere do young men look so old as in California," a local editor remarked in 1858.[28]

Yet the average miner was not a dissipated wastrel—despite the manner in which he spent his Sundays. Even Hinton Rowan Helper, who penned as sour a description of California as has ever been written, admitted that "taken as a body, they [the miners] are a plain, straightforward, hard-working set of men," [29] and Frank Marryat, a likable Englishman, warned his fellow countrymen that the California miners were not the picturesquely reckless, hairy creatures of contemporary foreign fancy. They were, he said, hardworking, skillful, and intelligent mechanics, engineers, and excavators who were putting a large part of their earnings back into the improvement of their claims.[30]

If the miners had any one universal sin, it was their unwillingness to remain in one spot for a reasonable period of time. They were forever abandoning one district and hurrying off to another at the first whisper

---

[27] John L. Comstock, *A History of the Precious Metals, from the Earliest Periods to the Present Time, with Directions for Testing their Purity* (Hartford, Connecticut, 1849), p. 218.

[28] North San Juan *Hydraulic Press*, October 30, 1858.

[29] Helper, *Land of Gold*, p. 153. Helper was, of course, the author of the famous *Impending Crisis of the South*.

[30] Marryat, *Mountains and Molehills*, pp. 314-316. The terminology is in the original.

that rich ground had been found. In extreme disgust the *Sacramento Transcript* once denounced the people of its state as "certainly the most migratory on the whole face of the earth." On another occasion it compared the shifting mass of miners to "a sea whose tide knows no law—a rumor or a reality will fill a canon with thousands of human beings." [31] A forty-niner explained this by saying:

There is an excitement connected with the pursuit of gold which renders one restless and uneasy—ever hoping to do something better. The very uncertainty of the employment increases this tendency. A person may be making his quarter ounce a day, and hears that a person a few miles from him is making an ounce. He is accordingly dissatisfied, and removes to the new diggings, there, probably, to be again disappointed.[32]

In amusingly worried terms Hinton Rowan Helper gave a similar explanation: "Once fairly started in a miner's life, I could not completely steel myself against the extravagant hopes which seemed to float in the very atmosphere of the mines. Wild and extravagant fancies would in spite of me obtrude themselves upon what I thought a well-balanced mind." [33]

Another reason for this eternal restlessness was the

[31] *Sacramento Transcript*, November 6, June 28, 1850; cf. Woods, *Sixteen Months*, pp. 68-69; Borthwick, *Three Years*, p. 136; Ingalls, *Journal of a Trip*, p. 47; Clappe, *Letters of Dame Shirley*, II, 28-29, dated Indian Bar, April 10, 1852.
[32] Woods, *Sixteen Months*, pp. 181-182.
[33] Helper, *Land of Gold*, p. 148.

hardship, monotony, and poor reward of mining life. The usual day began when one crawled out of his blankets, pulled on his dirty, patched trousers and his soggy boots, and turned to the day's tasks. First of all, breakfast had to be cooked, for the average miner was either his own housekeeper or a member of a group amongst whom the household chores were shared. Once that meal was out of the way, each man went to his claim to work with the pick and shovel, the long tom or sluice, as the case might be.

Often this involved toiling for hours in the glare of the burning sun, while cold, muddy water swirled about one's feet. Always it required some form of hard manual labor. Then, at noon, came the pause for lunch —"dinner"—and a companionable pipe, before resuming the program for the afternoon. At evening the miners dragged themselves back to their cabins for their uninspiring supper of warmed-over beans, potatoes, saleratus bread or flapjacks, tough beef or pork, and dried apples.[34]

Some miners found the preparation of their own food too disagreeable a labor, and hence made arrangements to have a local "hotel" or Chinese eating house provide them with three meals a day.[35] In either case the diet was

[34] The foregoing picture is drawn from: Borthwick, *Three Years*, pp. 123-126, 141-142; Helper, *Land of Gold*, pp. 155-158; Marryat, *Mountains and Molehills*, pp. 234-235, 255; Mulford, *Prentice Mulford's Story*, pp. 105-113; Woods, *Sixteen Months*, pp. 57-58; John H. Eagle, Auburn Ravine, California, to his wife, June 13, 1853, Eagle Correspondence.
[35] San Francisco *Daily Alta California*, May 15, 1852.

rarely an attractive one and often was unhealthy. During 1849 and a part of 1850 there was a dangerous shortage of fresh vegetables, fruit, and dairy products, with the result that many suffered from diarrhea, dysentery, scurvy, and other debilitating diseases.[36] The development of farming in California gradually supplied this lack. In 1854 a mining-town newspaper declared: "The extreme hardships of '49 and '50 have passed away—the days of 'flap-jacks' and 'pickle pork' are only remembered as a tale that is passed." [37]

Even with this improvement in diet, there remained many causes of discomfort and suffering, and the obvious question comes to mind: in return for so much hardship, how likely was the average miner to win a fortune? Level-headed contemporaries were in agreement in saying that the chances of great wealth were few. They pointed to the rarity of "great strikes," and to the working days lost because of Sundays, bad weather, sickness, and prospecting or traveling. All too often, they said, it was a struggle for existence rather than a grasping for fortune.[38] In deep discouragement one forty-niner noted in his diary:

[36] James L. Tyson, *Diary of a Physician in California, Being the Results of Actual Experience, Including Notes of the Journey by Land and Water* (New York, 1850), pp. 9-10, 61-64, 73-75, 78-79, 81; Kelly, *Excursion*, II, 37, 53, 151-153; Woods, *Sixteen Months*, p. 71.
[37] *Grass Valley Telegraph*, April 6, 1854.
[38] Kelly, *Excursion*, II, 268-269; Woods, *Sixteen Months*, pp. 14, 19-20, 169, 171-176; Borthwick, *Three Years*, p. 190; Shaw, *Ramblings in California*, p. 68; Cronise, *Natural Wealth*, p. 565; *Napa County Reporter*, December 31, 1859.

This morning, notwithstanding the rain, we were again at our work. We *must* work. In sunshine and rain, in warm and cold, in sickness and health, successful or not successful, early and late, it is work, *work*, *WORK! Work or perish!* All around us, above and below, on mountain side and stream, the rain falling fast upon them, are the miners at work—not for *gold*, but for *bread*. Lawyers, doctors, clergymen, farmers, soldiers, deserters, good and bad, from England, from America, from China, from the Islands, from every country but Russia and Japan—all, all at work at their cradles. From morning to night is heard the incessant rock, rock, rock! Over the whole mines, in streamlet, in creek, and in river, down torrent and through the valley, ever rushes on the muddy sediment from ten thousand busy rockers. Cheerful words are seldom heard, more seldom the boisterous shout and laugh which indicate success, and which, when heard, sink to a lower ebb the spirits of the unsuccessful. We have made 50 cents each.[39]

[39] Woods, *Sixteen Months*, pp. 102-103. The entry is dated Curtis's Creek, January 15, 1850.

THE LOUSY MINER.

It's four long years since I reached this land,
In search of gold among the rocks and sand;
    And yet I'm poor when the truth is told,
        I'm a lousy miner,
    I'm a lousy miner in search of shining gold.

I've lived on swine 'till I grunt and squeal,
No one can tell how my bowels feel,
    With slapjacks swimming round in bacon grease.
        I'm a lousy miner,
    I'm a lousy miner; when will my troubles cease?

I was covered with lice coming on the boat,
I threw away my fancy swallow-tailed coat,
    And now they crawl up and down my back;
        I'm a lousy miner,
    I'm a lousy miner, a pile is all I lack.

My sweetheart vowed she'd wait for me
'Till I returned; but don't you see
    She's married now, sure, so I am told,
        Left her lousy miner,
    Left her lousy miner, in search of shining gold.

CALIFORNIA GOLD

Oh, land of gold, you did me deceive,
And I intend in thee my bones to leave;
So farewell, home, now my friends grow cold,
I'm a lousy miner,
I'm a lousy miner in search of shining gold.

—from *Put's Original California Songster*

VII

THE THREE SECTIONS:
NATURE AND MAN, 1848-1860

In the preceding chapters the mining country has been discussed as if it were a single unit, because all parts of the mines possessed the same basic techniques and tools of the industry, and because mining camps and miners everywhere were fundamentally the same. Underneath this general uniformity, however, lay a considerable degree of local variation. During the early days of the Gold Rush it became customary among miners to speak of the districts tributary to Sacramento as the "Northern Mines," and of those tributary to Stockton as the "Southern Mines." At that time this was a sensible and easily understandable division of the auriferous region into two halves. Subsequently, during 1850-1852, a great development of mining took place at the head of the Sacramento Valley and in the rugged country beyond the valley, in the northwest corner of the state.

For want of inventiveness, popular usage simply expanded the meaning of the term "Northern Mines" so as to include the newer region. Thereafter this single phrase covered a vast area in which there were actually two quite different divisions. A more accurate system of terminology would have divided the mining territory into three sections: a "northwestern," by which would have been meant this newest region; a "central," which would have been the older districts formerly known as the Northern Mines; and a "southern."

Under such a classification scheme the "northwestern" section would have consisted of the diggings at the head of the Sacramento Valley, centering in Shasta City, and of a thin line of small camps that lay beyond the northern and northwestern walls of the valley. These camps were strung along the course of the Trinity, Klamath, and Scott rivers, and the numerous forks and tributaries of those streams.

The "central" section would have included El Dorado County, where Marshall made his great discovery, Amador County, just south of El Dorado, and the several tiers of famous mining counties which lay north of El Dorado: Placer, Nevada, Yuba and Sierra, Butte and Plumas. It would also have included Sacramento County, which was better known as a supply center than for its mineral wealth, and Sutter County, a block of lowland which sprawled across the strategic junction of the Sacramento and Feather rivers, on the route to Marysville.

Everything that remained would, perforce, be covered by the term "Southern Mines" or "southern section." The heart of this last section was the mining region in Calaveras, Tuolumne, Mariposa, and upper Stanislaus counties, but the section included also San

Joaquin County, in which Stockton was located. In 1852 it acquired a thinly settled outpost when gold washing began on the Fresno River, more than two dozen miles south of Mariposa. In 1855 it added a second isolated district, as the result of a sensational but short-lived rush to Kern River, which is at the extreme southern end of the San Joaquin Valley.

Of these three sections, the northernmost one was the first in size and the last in economic value. Geographically it stood apart from the other two, since they were parts of the Sierra Nevada system, whereas it belonged to the Klamath Mountains and the northern tip of the Coast Range. Its two dominant characteristics were ruggedness and remoteness. Almost the whole of its surface was covered by abrupt mountains and steeply cut valleys. Unlike the Sierras and the Coast Range, these mountains were not arranged according to a logical pattern, but rather had been tossed carelessly into place, as if by someone who reckoned not for the future convenience of mankind.[1]

Because of this irregularity and ruggedness, it proved difficult in the extreme to establish any communications within the section save by winding pack-trails. External communications, between the northwest and the outer world, were only slightly less of a problem. The section could be tapped from the Pacific Ocean by sailing from San Francisco up the foggy, treacherous north California

[1] Cf. p. 2.

coast. It could be reached from the interior by steaming up the Sacramento River for 150 miles, over shoals, sand bars, and snags, or by traveling overland along a course parallel to the river.[2] In either case, the journey between this section and the great commercial cities was exceptionally long and difficult, and yet for the miner or trader it was but the prelude to an even more trying trip within the region itself.

The problem of communications was rendered the more difficult by the comparative severity of the climate of the northwest. A contemporary said of it: "The climate is rigorous, the winter rains being long continued and heavy in the valleys, and the snow lying to a great depth for several months on the mountains."[3]

As if these physiographic obstacles were not enough, still another stumbling block was provided by the presence of hostile Indians. In many of the mining districts of the other two sections the Indians were an occasional local problem between 1848 and 1852, but in only a few was there any real danger after that period. In the northwest, on the other hand, the aborigines were a more competent crew than elsewhere in the state, and by utilizing the very difficult terrain of their native habitat they were able to remain in the field, as a real though sporadic

[2] San Francisco *Daily Alta California*, November 11, 12, 1851, July 28, 1855; *Sacramento Weekly Union*, August 16, 23, 1856; *Sacramento Daily Union*, November 6, 1855.

[3] Cronise, *Natural Wealth*, p. 568.

menace, long after their brethren to the south had succumbed to those twin agents of white conquest: firearms and firewater. Their power was not broken until the federal government sent militia and regular troops against them during the American Civil War.[4]

In partial compensation for these hindrances to development, nature endowed the section with resources that were of considerable value. Gold, both in placer deposits and in veins, was scattered along many of the creeks and rivers, and in some of the gulches and hillsides. The rainfall was the heaviest in California, and thanks to it, water and timber—two elements essential for mining—were abundant in the greater part of the northwest. Given these same resources in an area that was more accessible and less troubled by Indians, and progress might have been rapid.

In the Shasta part of the section gold was discovered as early as 1848, and the season of 1849 saw prospecting and mining at a number of points in both Shasta County and the adjacent Trinity mines. No real rush developed until 1850. Beginning in 1850 and continuing for the next three years, the newspapers reported a considerable excitement over the new area. Miners defied Indians and natural environment alike, in order to set up camps along the rivers, while optimistic promoters established villages

---

[4] Anthony J. Bledsoe, *Indian Wars of the Northwest. A California Sketch* (San Francisco, 1885), *passim.*

at several points strategically located to control the flow of trade.[5]

Despite this boom, two of the leading newspapers in the state declared in May 1852 that few people in the heart of California had any clear idea of the nature of the far northern country and its resources.[6] Three years later, in May 1855, one of these journals repeated its earlier statement, and added the remark that in all probability the majority of its readers did not even know the exact location of Trinity County, which was one of the most important northwestern subdivisions.[7]

The press, in 1855-56, was in agreement in saying that this lack of interest on the part of the outer world was reflected by a notable sparseness of population within the section itself. The main handicaps cited were the familiar ones of isolation and Indian troubles. On the other hand, by the very fact of having so few people in so large a territory, the northwest was able to offer unexploited, virgin ground at a time when the men of the older sections were reworking the same diggings for the sixth or seventh time. Long after simple placer mining had passed its zenith elsewhere, the newspapers were still

[5] See map in San Francisco *Weekly Pacific News* (steamer ed.), August 15, 1850. See also: San Francisco *Weekly Alta California*, May 18-June 1, 1850, January 25, February 15, March 8, 15, 1851; San Francisco *Daily Alta California*, January 9-13, September 19, 1851, May 24, 1852.

[6] San Francisco *Daily Alta California*, May 19, 1852; *Sacramento Weekly Union*, May 8, 1852.

[7] *Sacramento Daily Union*, May 2, 1855.

reporting fine yields and untouched auriferous soil.[8]

Partly because virgin ground was still available, the miners of the northwest were slow to resort to improved methods. From time to time a loyal local newspaper claimed great mining achievements for its readers, but the consensus of the less biased observers was that mining techniques there were far behind those of the other two sections. The rocker, for example, was still in use on the Klamath River at the end of 1856, a half-dozen years after it had been supplanted elsewhere by the long tom and sluice. The building of canals and flumes for conveying water was inexplicably slow in becoming a common practice, especially when one considers the section's superior water and timber resources.[9]

The area was thus throughout the fifties a retarded mining frontier occupied by a sparse population of men of little wealth and inferior initiative. Apparently it was rich enough to furnish a living with primitive modes of mining, but not rich enough to attract to it men of substantial capital or outstanding ability.

The slow rate of development of the northwest in-

[8] *Ibid.*, May 2, August 29, October 4, November 6, 13, 1855, September 16, 21, 1857; Arcata *Northern Californian*, April 4, 18, May 2, 23, June 6, 13, 20, 1860; "Placer Mining Summary," *California Mining Journal*, I (1856-57), 21, 29.

[9] Sacramento *Democratic State Journal. Weekly*, July 2, 1853; *Sacramento Daily Union*, November 13, 1855, September 16, 21, 1857; *Sacramento Weekly Union*, January 31, 1857; *Mining and Scientific Press*, October 5, 1861.

tensified the exclusiveness which nature had imposed upon that area by the hard facts of geography. The people of the time accepted it as a part of the "Northern Mines," but when they talked of the rivalry between the "Northern" and "Southern" mines, they had principally in mind the "central" and "southern" sections. The latter two had much in common: both lay along the Sierras' western flank; both had an active history that extended back to the early summer of 1848; both were served commercially by the great inland waterway that began at the Golden Gate and stretched past San Francisco and up the Sacramento and San Joaquin rivers to the queen cities of Sacramento, Stockton, and Marysville.

Behind these similarities, however, lay a fundamental inequality in natural endowment. Nature had given both sections gold deposits and the water and timber with which to attack them, but she had not distributed her gifts in like proportions between the two. The inequality was not especially apparent during the first three seasons of 1848, 1849, and 1850. In those flush years all attention was focused upon the most easily worked deposits. This generally meant shallow placers, along the course of present-day streams and gulches. Both sections were rich in this type of "diggings."

When signs of exhaustion began to appear, miners were forced to try less obvious places. In Nevada County this led to "prospecting" the gravel hills that stood above Nevada City. To the surprise of the skeptical, they were found to be rich in gold. Similar hills in the neighborhood were attacked at once, and with like results.[10] Success in Nevada County led to investigations elsewhere, and before long it had become apparent that a new and important kind of placer deposit had been discovered.

To this kind of deposit geologists have given the name of Tertiary gravels, although contemporaries usually spoke of it as "deep gravels." The Tertiary gravels are usually found in thick masses or beds. They were formed in the Tertiary period of geologic history, through the erosion of gold veins by rivers that have since been greatly altered or have ceased entirely to exist. These ancient streams laid down their flakes and fragments of gold in beds of gravel along their course, precisely as do present-day streams. On top of the auriferous matter, thick layers of less valuable or worthless debris were subsequently deposited.

Then came a period of mountain building that elevated the Sierra Nevada range and drastically changed the drainage pattern of its western slope. Responding to the new topography, the rivers of that epoch transferred their lines of drainage from their old valleys to their present ones. Throughout the long interval since the close of Tertiary time, the modern streams have been

[10] Rolfe, "Mines and Mining," *Bean's History*, p. 65; but cf. C. W. Haskins, *The Argonauts of California* (New York, 1890), pp. 163-164, which claims the discovery for El Dorado County.

at work, cutting deep the canyons and valleys in which one finds them today. On the ridges between the modern canyons they have left the rich Tertiary auriferous gravels, together with the thick layers of debris that so often cover them.[11]

The discovery of the deep gravels dates from the early part of 1850, although its full significance was not realized for some time thereafter. It was not a discovery in which the central and southern mining sections shared equally. The northerly counties of Sierra, Yuba, Butte, and Placer all contained rich Tertiary beds, thanks to the erosive action of the ancient streams upon their local quartz veins. In El Dorado County, just below Placer, the Mother Lode begins, and in this one county there were preserved deposits formed by erosion of the Mother Lode veins.[12]

South of El Dorado the continuation of the Mother Lode veins runs through one more county of the central section, Amador, and through almost the whole of the Southern Mines. In Amador County and the Southern Mines, "it so happens that most of the ancient river deposits below the great Mother Lode are either eroded [away] or so heavily covered [with subsequent debris] that they can not be mined." [13]

---

[11] Lindgren, *Mineral Deposits*, pp. 229-231.

[12] Waldemar Lindgren, "The Tertiary Gravels of the Sierra Nevada of California," U. S. Geological Survey, *Professional Paper*, no. 73 (Washington, 1911), p. 65.

[13] *Ibid.*, pp. 65-66.

Through being deprived of "this great source of enrichment," the Southern Mines as a whole lost the opportunity to participate in the most important type of placer mining that developed in California after 1851. There were, to be sure, a few notable local exceptions. In Tuolumne County a geologic "fluke" happened to cover the drainage channel of an ancient stream with a basaltic flow that withstood all subsequent erosion. Through the degradation of the surrounding country, this basaltic flow stood forth in later times as a flat-topped eminence to which was given the appropriate name of Table Mountain. Miners found it possible to penetrate into the gold deposits preserved underneath this basalt cap.[14]

Other strokes of geologic fortune preserved a few additional patches of Tertiary gravel, such as those near Chinese Camp, which is also in Tuolumne County. More important was the curious geologic history of Columbia, a mining town about four miles from Sonora. Columbia stood in the center of a flat, open valley that was about two miles in diameter. This little basin happened to survive from Tertiary time down to the present with almost the same topographic features that it had in the pre-volcanic era. Because of its flatness and because it was underlain by limestone that was deeply pitted with retentive potholes, it served as a splendid natural receptacle to catch the gold fragments that were carried down

[14] *Ibid.*, p. 66.

into it during the erosion of the surrounding country-side. In this way Columbia became "one of the richest districts in the Sierra Nevada. It is said that over $50,-000,000 was taken out from the Columbia diggings from 1853 to 1870." [15]

With these few exceptions, the working of the Tertiary deposits proved to be almost exclusively a prerogative of the Northern Mines. A similar but not so drastic northern superiority was revealed when miners began to attack quite a different type of gold deposit.

In the summer of 1849 gold hunters in Mariposa County, at the southern end of the Mother Lode, tried tracing a line of auriferous rock fragments back to their source. The experiment led them up a hillside to the vein from which the nearby placer deposits had been eroded. There, in that same summer, the first vein, or quartz mining in California was attempted.[16]

Similar discoveries of vein gold soon followed in many parts of both the Northern and Southern Mines. During the early fifties quartz mining boomed into the public notice as a major branch of the gold industry, and during the years that followed it proved to be the most permanent of all the forms of mining.

Since the original discovery of quartz gold was made

[15] *Ibid.*, pp. 212-213.
[16] Carl E. Julihn and Frederick W. Horton, "Mineral Industries Survey of the United States: California. Tuolumne and Mariposa Counties, Mother Lode District (South). Mines of the Southern Mother Lode Region. Part II," U. S. Bureau of Mines, *Bulletin*, no. 424 (Washington, 1940), pp. 3, 94.

in the Southern Mines, and since that section possessed well over one-half of the entire length of the Mother Lode system of auriferous veins, a southern supremacy in this department might have been expected. The expectation might have seemed the more justifiable in view of the fact that the Southern Mines also had a second, though less important, belt of auriferous veins in addition to the Mother Lode. This second belt, known as the East Belt, ran from Amador County to Mariposa along a course parallel to the Mother Lode and on the eastern side of it.[17]

Despite these favoring factors, the central section soon surpassed the southern in quartz. The most valuable veins in the state were found to be not in the famous Mother Lode belt at all, but rather in Nevada County.[18] Within the limits of the Mother Lode, "the most productive portion of the belt" was shown to be the ten-mile strip that spans Amador County.[19] As for the East Belt, the segment of it near the town of Sonora had a flurry of prosperity from 1858 to the middle sixties, but thereafter it lapsed into secondary importance.[20] Accord-

[17] California Miners' Association, *California Mines and Minerals* (San Francisco, 1899), pp. 320, 341-345, 352-354, 362, 366-367.

[18] *Ibid.*, pp. 11-12; Waldemar Lindgren, "The Gold-Quartz Veins of Nevada City and Grass Valley Districts, California," U. S. Geological Survey, Seventeenth Annual Report, part II, in Secretary of the Interior, *Report*, IV (Washington, 1896), 112.

[19] Knopf, "Mother Lode System," U.S.G.S., *Professional Paper*, no. 157, p. 23.

[20] Sonora *Union Democrat*, June 2, 1860; *Sacramento Daily Union*, April 22, May 10, 18, 1858; J. Ross Browne, *Report on the Mineral Re-*

ingly, while quartz mining did much to develop a few of the districts of the Southern Mines, it did not bring to the section as a whole as much benefit as it did to the central section.

The central section's superiority in mineral wealth would have been less conclusive if it had not been accompanied by an advantage in regard to water, which was the most important auxiliary element required for mining. In California the amount of rainfall is determined by a combination of altitude and latitude. Precipitation is greatest in the latitudes of the remote northwest and in the loftiest parts of the Sierras. It decreases as one travels southeastward parallel to the axis of the state. It diminishes far more rapidly if one descends from the High Sierra through the lower ranges and foothills to the floor of the Great Valley. Modern records for the area between the Yuba and Tuolumne rivers—the heart of the mineral region—indicate that rainfall is reduced by 8½ inches with each 1,000 foot drop in altitude.[21]

Now most of the mining in California in 1848 and 1849 was done at comparatively low altitudes—between 1,000 and 2,000 feet—because that was where the present-day streams naturally tended to drop their load of

sources of the States and Territories West of the Rocky Mountains (Washington, 1868), pp. 45-48.
[21] Alexander G. McAdie, "The Rainfall of California," University of California Publications in Geography, I (1913-1917), 150-159. True only below the 5,000 foot level.

gold. Thereafter, coincident with the discovery of the deep gravels and, less significantly, with the boom in quartz, men in the central section found increasingly that some of the most profitable fields lay well above the former scenes of activity.

Two whole counties, Plumas and Sierra, came into existence primarily because of the miners' rush into the higher country in 1850-51 in search of Tertiary deposits. Similarly, the development of the rich deep gravel and quartz claims in the upper part of Nevada County began only in 1850 and 1851, while in the corresponding section of Placer County important deep gravel towns like Dutch Flat, Gold Run, and Iowa Hill did not exist prior to that date.

In the Southern Mines there was no comparable advance into higher regions in 1850-51 or in subsequent years. Mining continued to be restricted primarily to the Mother Lode belt in which it had begun, and it should be remembered that that belt has an altitude of 2,700 feet at its northern end, but of only 2,000 feet at its southern extremity.

This factor of altitude had the direct result of giving the mining camps of the central section a better supply of rainfall than that of their southern neighbors. Indirectly it aided them still further, for the amount of water in the streams is influenced not only by local precipitation but also by the ability of the streams to tap the great reservoir of the Sierran snowfields during the summer months. While localized topographic features were sometimes so dominant as to nullify other factors, in general the superior altitude of the camps in the central section meant that the streams which served them were closer to this natural reserve and therefore able to deliver a larger quantity of water.

Moreover, the advantage in regard to rainfall carried with it a corollary in regard to timber, which was an essential material for all save the most elementary mining operations. Although the Southern Mines were by no means devoid of forests, it was nevertheless true that the better stands of commercially valuable timber tended to coincide with the belts of greater rainfall.[22]

In basic natural resources, then, the central section had an advantage over the southern. It had almost a monopoly over the Tertiary gravels, it possessed the most productive of the quartz mining districts, it had a larger natural supply of water, and it came off somewhat better in regard to forests. There remained only one further sphere of rivalry: the number and character of the people who were to make up the population of the two areas.

Here also the central section was consistently ahead of all rivals. The federal censuses of 1850 and 1860 and the state census of 1852 all showed the central section as leading, the southern as coming second, and the north-

[22] George B. Sudworth, *Forest Trees of the Pacific Slope* (Washington, 1908), *passim*.

western as a weak third. In 1850 almost the entire population of the mineral region was listed as living in either the central or southern, with the former having between two-fifths and three-fifths of the total, and the latter slightly less than two-fifths. By the time of the state census, the central section was approximately twice as populous as the southern, even though the latter had added largely to its population. The northwest, in the meantime, had begun its first boom and had grown four-fold, but it still had less than one-fourteenth as many people as the central area.

The census of 1860 showed that the rate of increase had declined in all three sections. The central section, however, had maintained its growth better than the southern, and was now two and one-quarter times as populous. The northwest, despite almost a three-fold gain since 1852, was only one-fifth or one-sixth as large as the central section, and considerably less than one-half as large as the southern.[23]

These figures show that the population inferiority of the Southern Mines was least at the start of the fifties and greatest at the close of the decade. This relationship was closely connected with the change that came over California mining during the middle and later fifties, when the decline in the yield of the rich, shallow placers forced men to turn to the deep gravels and quartz veins.

[23] Figures for 1850 and 1852 from De Bow, *Statistical View*, pp. 200-201, 394; and for 1860 from *Eighth Census, 1860: Population*, pp. 28, 33.

There was a similar correlation between the greater attractiveness of the central district's resources and the national, or racial, composition of the population of the respective sections. According to the census of 1850, three of the four counties of the Southern Mines had an unusually large non-American element. In one, Tuolumne, the foreign-born actually outnumbered those born in the United States. In the other two, Calaveras and San Joaquin, the foreign-born were about a third of the population. None of the counties of the Northern Mines had so large a percentage of aliens.

Two years later the state census showed that the high ratio of foreigners continued in the Southern Mines. The returns revealed that in Tuolumne and Calaveras the "foreign residents" formed over one-half of the total population. No other county in any section had a foreign majority.

A contemporary description, written by a Scotch artist-miner who was in California from 1851 to 1854, confirms the evidence of the census figures:

In the north, one saw occasionally some straggling Frenchmen and other European foreigners, here and there a party of Chinamen, and a few Mexicans engaged in driving mules, but the total number of foreigners was very small: the population was almost entirely composed of Americans, and of these the Missourians and other western men formed a large proportion.

The southern mines, however, were full of all sorts of people. There were many villages peopled nearly altogether by Mexicans, others by Frenchmen; in some places there were parties of two or three hundred Chilians forming a community of their own. The Chinese camps were very numerous; and besides all such distinct colonies of foreigners, every town of the southern mines contained a very large foreign population.[24]

By 1860 the number of persons of alien origin had increased greatly in all parts of the mines. In almost all of the counties in all three sections, persons of foreign birth formed at least one-third of the population. In the Southern Mines at that period, mining was largely confined to the three main counties of Tuolumne, Calaveras, and Mariposa. The foreign-born were in the majority in all three. In the central section there were by then eight active mining counties. The non-American element had a majority in one, and in another was equal to the native-born. In the northwest, two of the five counties had extraordinarily high percentages of foreign-born, amounting to three-fifths in the case of Trinity and two-thirds in Klamath.

Two conclusions are obvious: first, that the native Americans tended to concentrate in the richest section; and second, that after the passing of the early flush years, one-half of the mineral region was more foreign than "American."

It would be wrong to picture this as being entirely the result of the influence of natural factors upon human desires. There were important contributing factors of sheer chance. For reasons best known to himself, John

[24] Borthwick, *Three Years*, p. 306.

A. Sutter decided to establish his fort near the junction of the Sacramento and American rivers. New Helvetia and the Sacramento Valley, rather than Weber's little settlement and the San Joaquin, thereby became the chief arena for American and European activity prior to the great events at Coloma. Coloma, in turn, was so good an advertisement for the wealth of the American River that that stream became the most important mining center of the early Gold Rush days.

When the stampede began in 1848, Americans formed the majority of the men who began mining during that first summer, even though they were but a minority of the persons who were then in the province. Quite naturally, they tended to cluster about the streams that were within easy traveling distance of Coloma. Then immigrants began hurrying into California by sea and by land. If they came by sea, and landed at San Francisco, it was an open question whether they should proceed to the central or to the southern section. If they came by land, on the other hand, their destination was largely determined for them by the direction from which they were approaching. The route from Oregon led naturally into the Northern and that from Mexico into the Southern Mines. Most of the wagon trails across the Great Plains debouched into the central section.

For this reason, to the Americans already in the central section was added the bulk of the American overland immigration, while most of the Mexicans went to the Southern Mines. There the Sonorans not only gave their name to the leading mining town of the south, but also made Sonora their headquarters in California.

In the Southern Mines the Mexicans were joined by men from the four corners of the earth. Sonora became an international center, with most of the vices and not many of the virtues generally attributed to cosmopolitan communities. A traveling correspondent for the *Alta California* said of it, in 1850:

Sonora is nearly as large as Stockton, and far ahead of it for gold, gals, music, gambling, spreeing, etc. It's a fast place, and no mistake. Every Sunday there is either a horse race or a bull bait, and any number of fights and rows. Such a motley collection of Mexicans, Chilians, Frenchmen, Chinese, Jews, Jonathans, Paddies, and Sawnies, I have never seen together before in California.[25]

A widespread anti-foreign agitation in 1849 and the early fifties did something to purge the Southern as well as the Northern Mines of aliens. In particular, it caused many of the Latin Americans to quit California forever.[26] In the Southern Mines, however, an Old Guard of Mexicans obstinately lingered on, and was reinforced by the arrival of many Frenchmen and other Europeans. Throughout the fifties the international flavor remained strong in Sonora,[27] and the census statistics are sufficient

[25] San Francisco *Weekly Alta California,* June 1, 1850.
[26] San Francisco *Alta California* (steamer ed.), August 2, 1849; San Francisco *Weekly Alta California,* August 10, 1850.
[27] John Heckendorn and W. A. Wilson, comps., *Miners & Business Men's Directory, for the Year Commencing January 1st, 1856* (Columbia, California, 1856), p. 37.

proof that Sonora was not the only southern town to have this characteristic. At as late a date as 1861, a visitor to Mariposa County said of the town of Hornitos:

The town is certainly of Spanish origin, and even to this day there seems to be an omnipresent struggle between the Mexican and American element. . . . This rivalry, if we may use the term, is visible in everything; it is an even chance as to whether the next passer-by will be an American or Mexican. . . . Even the very signs seem to fight it out, or compromise. The stage house is the "Progresso Restaurant"; the bakery is a "panderia"; the hotels invite both in Spanish and English; the stores in Italian as well as American and Spanish; while Sam Sing or Too Chang outrival the "lavado y planhado." In the plaza Brother Jonathan, however, has it pretty much all to himself, and manifest destiny will, undoubtedly, prevail in the end.[28]

During the early days of the Gold Rush, the Southern Mines benefited greatly from the presence of the foreigners. Especially did it derive advantage from the technical skill of the veteran miners from Sonora province. As the *Alta California* remarked, "American energy and assiduity, and Mexican skill and experience have together developed the riches of the Southern Placer." [29]

In the opinion of some of the newspapers of the time, the natural environment of the Southern Mines was such that the section could ill afford to lose the services of the

[28] [Henry S. Brooks], "Hornitos, Quartzburgh and the Washington Vein, Mariposa County," *California Mountaineer*, I (1861), 335.
[29] San Francisco *Weekly Alta California*, August 10, 1850.

Mexicans. Because of its lesser rainfall and the smaller size of its streams, the southern section early acquired and long retained the reputation of being the home of "dry diggings." In mining parlance, "dry diggings" were those that had no natural supply of water save during the wettest part of the year, when the rains created short-lived streams.[30]

The yield of the dry diggings during the late winter months was often large, but the period of effective operation was short, and the chances of success were entirely dependent upon the weather. During the remainder of the year, only the most limited type of mining was possible. It was the belief of the *Stockton Times* that:

The Mexican is of the utmost service in the Southern mines. We ask those who have had actual experience in mining operations in this country, whether the American, with all his impatience of control, his impetuous temperament, his ambitious yearning, will ever be content to deny himself the pleasures of civilized life in the States, and for the sake of from four to eight dollars per day, be content to develop the resources of the dry diggings of the country.[31]

There was a means of overcoming the disadvantages of the dry diggings, and that was to bring water to them through artificial channels. That was the way in which

[30] San Francisco *Daily Alta California*, January 21, 1852; *Sacramento Weekly Union*, January 21, 1854.
[31] San Francisco *Weekly Alta California*, August 24, 1850, clipping *Stockton Times*.

the central section developed its share of the dry placers. But canals, flumes, and impounding dams were a difficult business, requiring engineering skill, capital investment, and associated labor. Apparently the south was never able to summon to its aid human and financial resources comparable to those of its northern neighbor. In 1852 the *Stockton Journal* was complaining: "In the South there is room for a vast population in the dry placers, if water could be procured." [32] Three and one-half years later another Stockton newspaper admitted: "The want of water during the dry season has been, since the commencement of mining, more seriously felt in the southern than in the northern mines, which are more abundantly supplied with ditches and canals." [33]

One must conclude that the Southern Mines, with their large foreign element and their smaller total population, found themselves at a disadvantage when they had to call upon their people for something more than simple labor and elementary mining techniques. Doubtless the same explanation would hold true in part for the northwest, and in both cases the character and size of the population were in large part a reflection of the degree of attractiveness of the natural endowment.

This should not, however, lead to an iron-bound doctrine of natural predestination. The eventual supremacy

[32] San Francisco *Daily Alta California*, March 6, 1852, citing *Stockton Journal*.
[33] *Sacramento Daily Union*, September 27, 1855, clipping Stockton *San Joaquin Republican*.

of the central section was inevitable, but the rapidity with which it was actually achieved was not so much the result of natural forces, as of the decision of thousands of individual miners to follow the crowd and go to the area to which sheer chance had given the honor of un-covering the first gold deposits.

VIII

BASIC CONDITIONS FOR A TRANSITIONAL ERA

All new mining countries seem to start their life with a fanfare of flush times and universal optimism. Usually this prosperity is based upon the presence of rich, virgin deposits that can be exploited by comparatively simple methods. Because in minerals nature gives but does not renew, these deposits presently begin to show signs of exhaustion. The mining country then finds itself facing either one of two fates: decay or transition.

If there are no deeper resources, and if the capabilities of the population are limited, the former condition must prevail. If, however, both the territory and its people have latent reserves, then there is a good chance that the region may yet live to ripe maturity, provided always that it is sturdy enough to survive a long and trying period of transition.

For California the fates planned the second and more attractive of the two destinies, and they decreed that the era of change should begin in 1851. In that year the leading newspaper in the state declared:

The real truth is, by far the largest part of the gold . . . [mined hitherto] was taken from the river banks, with comparatively little labor. There is gold still in those banks, but they will never yield as they have yielded. The cream of the gulches, wherever water could be got, has also been taken off. We have now the river bottoms and the quartz veins; but to get the gold from them, we must employ gold. The man who lives upon his labor from day to day, must hereafter be employed by the man who has in his possession accumulated labor, or money, the representative of labor.[1]

In other words, the palmy days of mining were passing. The easily worked deposits were beginning to give out. For another year or two they might be induced to supply a good yield, under the persuasion of more efficient mining implements and a larger mining population, but thereafter the burden of maintaining the state's annual output would fall upon deposits that could be attacked only by difficult methods. Henceforth the miner must be prepared to undertake projects that demanded the same sort of skill that would be required for a large-scale engineering operation. He must be prepared to buy and use expensive machinery, and to erect costly structures. If he did not have the ability to do so, he must reconcile himself to the status of hired labor.

This change was reflected in the statistics of the amount of gold mined in California in each year. From 1848 through 1851 the annual increase was very great. By 1852 the rate of gain was slowing down perceptibly —a circumstance which indicated that production had reached its peak. In 1853 a sharp decline was recorded.

[1] San Francisco *Weekly Alta California*, February 15, 1851.

A year later there was a partial recovery. Then the output plunged abruptly into a permanent decline.

Fiscal Year	Gold Produced in California
1848	$ 245,301
1849	10,151,360
1850	41,273,106
1851	75,938,232
1852	81,294,700
1853	67,613,487
1854	69,433,931
1855	55,485,395
1856	57,509,411
1857	43,628,172
1858	46,591,140
1859	45,846,599
1860	44,095,163[2]

It was the task of the individual miner to combat this decline by increasing the effectiveness of his exploitation of the more difficult deposits. In order to do so he had to undertake operations that required the employment of hired labor. During the flush days that was well-nigh impossible, for in 1848 and 1849 few men in the mining districts would accept employment at any price. Thereafter, as the population grew and the income of the inde-

[2] See appendix for discussion of statistics.

pendent miner waned, men became more willing to "hire out" to their more fortunate or more able neighbors. In the Southern Mines, Latin Americans were especially apt to fall into hired status.[3]

Prior to 1850 or 1851, when a miner talked of his "wages" he meant his daily income from his own mining operations.[4] After that he might mean either the gold he mined on his own account, or a sum that he was being paid for working for others. In general, the governor that regulated the rates of all work for hire was the number of dollars popularly believed to be the prevailing daily average "take" of the individual miner.[5] In practice, however, the relationship between the independent "wage" and the hired "wage" was greatly modified by local and personal factors.

By piecing together the scattered reports contained in both the published and unpublished records of the time, it is possible to form a rough idea of the progress of wages from 1848 to 1860. It seems clear that fabulously high wages ceased after the first three or four seasons. In the Southern Mines, where rates of pay seem to have

[3] P. Laur, *De la Production des Métaux Précieux en Californie. Rapport à S. Exc. M. le Ministre des Travaux Publics* (Paris, 1862), p. 53; Hittell, *Mining in the Pacific States*, p. 38; *Sacramento Transcript*, May 29, 1851; San Francisco *Daily Alta California*, September 12, 1851.

[4] Henry George, *Progress and Poverty, An Inquiry into the Cause of Industrial Depressions and of Increase of Want with Increase of Wealth, The Remedy* (50th Anniversary ed., New York, 1939), p. 33.

[5] *Sacramento Transcript*, April 12, 1850; *Scientific Press*, July 8, 1871; San Francisco *Daily Alta California*, August 6, 1869.

been lower than in the central section, there was talk of a five dollar daily wage as early as the autumn of 1851 and the spring of 1852. Over the mines as a whole, a five dollar rate became common soon afterward, and from that point the level dropped slowly downward to three dollars.

Year	Daily Wage
1848	$20
1849	16
1850	10
1851	8—
1852	6
1853	5
1856-58	3+
1859	3
1860	3—[6]

If judged by Eastern standards, even $3.00 was a large sum for a day's work. Eastern coal and iron mine operators were paying their employees at rates that averaged less than $1.25 per day, between 1848 and 1860, and in many Eastern areas even the most skilled miners received not more than $1.00 per day. There seems to have been only one recorded Eastern coal or iron district in which the miners ever had an opportunity to earn as much as

[6] See appendix for discussion of statistics. "Daily wage" means wage without board in the case of hired labor, and gross daily yield in the case of independent labor. The figures do not apply to Chinese.

$2.00 per day, whereas $2.00 was the lowest point to which California mining wages sank at any time or place during the fifties.[7]

From the working man's point of view, high wages were justified by the high cost of living in the California mining regions. Prices for goods of all kinds were at their peak in 1849, when supplies were from 100 to 300 or 400 per cent more in the mines than they were even at San Francisco.[8] By the latter part of 1850 charges had fallen considerably, and by 1852 food, at least, was obtainable at comparatively reasonable rates.[9] Most articles of clothing and household use, however, continued to be available only by importation, and according to a speaker at San Francisco's first industrial fair, "Experience proves that but few articles can be imported from the Atlantic States and sold profitably, at less than fifty per cent advance." [10] At as late a date as 1861 the leading California writer on economic subjects declared

[7] "History of Wages in the United States from Colonial Times to 1928, Revision of Bulletin No. 499, with Supplement, 1929-1933," U. S. Bureau of Labor Statistics, *Bulletin*, no. 604 (Washington, 1934), pp. 330, 333; Edith Abbott, "The Wages of Unskilled Labor in the United States, 1850-1900," *Journal of Political Economy*, XIII (1904-05), 357, 364; Nelson W. Aldrich, "Wholesale Prices, Wages, and Transportation," *Senate Report*, 52 Cong., 2 sess., no. 1394 (March 3, 1893), part 4, pp. 1561, 1565-1569.

[8] Felix P. Wierzbicki, *California As It Is & As It May Be, Or a Guide to the Gold Region*, ed. by George D. Lyman, Rare Americana Series, ed. by Douglas S. Watson, no. 8 (San Francisco, 1933), pp. 47-48.

[9] *Sacramento Transcript*, October 14, 1850; San Francisco *Daily Alta California*, May 15, 1852; *Sacramento Weekly Union*, August 28, 1852.

[10] *Sacramento Daily Union*, September 10, 1857.

that "the cost of living in the mines of California is about twice as much as in the Eastern states." [11]

On the other hand, the California employer also had to face these high prices for necessities, and in addition, whenever he needed capital or credit he had to meet interest charges that were proportionately as high as the wages of labor. Even in 1850 one hears of an interest rate of 12 per cent a month, and that was a distinct reduction from 1849.[12] In 1851 the San Francisco rates on "first class paper" and "undoubted security" were as high as 4 to 8 per cent a month during some parts of the year, and they never fell below 2½ per cent. During the remainder of the fifties, interest on urban real and personal property was occasionally as low as 1 to 2 per cent a month, but often ranged upward to 3 per cent.[13] When one considers that these quotations were for money secured on strong city collateral, it becomes evident that the promoter of so speculative an enterprise as a gold mine had to pay very dearly for any aid he received from capitalists.

These twin factors of high wages and high interest

[11] Hittell, *Mining in the Pacific States*, p. 212. Confirmed by *New-York Daily Tribune*, March 14, 1859.

[12] Laura A. White, "The United States in the 1850's as Seen by British Consuls," *Mississippi Valley Historical Review*, XIX (1932-33), 517, quoting consul at San Francisco.

[13] Based on files of: San Francisco *Merchants' Exchange Prices Current and Shipping List*, 1850-51; *Sloat's San Francisco Prices Current and Shipping List*, 1851-52; San Francisco *Mercantile Gazette and Shipping Register*, 1857; San Francisco *Mercantile Gazette and Prices Current, Shipping List and Register*, 1860.

were the omnipresent background for all California mining operations. They explain why the practice of associated labor was so universally resorted to in river damming projects: by combining their efforts the projectors could avoid hiring and avoid borrowing. They explain the eagerness to make machines and water do the work of men in the exploitation of the deep gravels. They explain why so many quartz mines were developed "on a shoestring," and, lastly, they explain why mining operators of all kinds were so often willing to sacrifice both permanency and efficiency in order to secure a quick return on their investment.

IX

TOWARDS A NEW ECONOMY: 1851-1860
I. RIVERS AND VEINS

After the passing of the flush days of 1848-1851, the chief interest in the history of California mining shifts to the exploitation of the river beds, deep gravels, and quartz veins. This does not mean that there was a sudden end to the older and simpler attempts to work the bars, banks, and gulches. On the contrary, for some time after 1851 the successive improvements in mining implements enabled the less ambitious of the miners to scratch out a living by reworking the same ground year after year.[1] This species of mining was of declining importance, however, and by the latter fifties it had been largely abandoned to the humble and patient Chinese, except in the northwest and some of the more backward parts of the south.

Over the state as a whole the exploitation of auriferous ground advanced with bolder steps. Some of its most notable accomplishments were in river mining, by which is meant the use of dams, ditches, and flumes to divert streams from their natural beds. This was a highly speculative business, since men's plans had to be governed by the seasons. In California the rain ceases in

[1] *Sacramento Weekly Union*, April 16, 1853.

May, but until the latter part of June or the first of July the water in the rivers remains high. Then come a few months of low water, before the return of the rains in November or December.

For the river miner this seasonal division meant a maximum working period of nearly five months, and in the best years the maximum was sometimes realized. Not infrequently, however, the length of the low-water interval was disastrously curtailed by unexpected showers in October or even in September. Because of the immediacy of the run-off from the sun-baked hills, a comparatively brief term of rain was often sufficient to convert a controllable stream into a torrent.

The miners never attempted to build dams, flumes, and canals strong enough to withstand the floods. Each year they began afresh in late June or early July and spent the summer in construction work. By September they would have completed the diversion of the river, and would be ready to commence digging gold from the exposed bed. For the next few weeks they would receive income—if the stream bed did not prove barren. Then the coming of the first rains would bring destruction to all of the works that had been erected.[2]

The whole success of the enterprise was thus dependent upon two unpredictable factors: the length of the low-water season and the richness of the ground

[2] Hittell, *Mining in the Pacific States*, p. 149; Cronise, *Natural Wealth*, p. 537.

hidden beneath the waters of the stream. On the one hand, early rains sometimes wiped out the whole project just as the miners were about to obtain their first returns, with the result that "many men are thus left penniless, after the toil and hope of a long and scorching summer." [3] On the other hand, miners frequently completed the diversion of the river, only to discover that the ground thereby exposed contained but a scattering of gold. In the opinion of the editor of a famous contemporary California magazine, "Taking the losses with the gain, it is very questionable if more gold has not actually been invested in river mining, than has ever been taken out." [4]

Despite the failures, river mining was one of the most common forms of large-scale operations. Its place of origin and its most important center, in the early days, was on the several forks of the American River, the stream upon which Marshall's original gold discovery was made. There, so many different companies went to work, one directly below the other along the river's course, that during the weeks of maximum effectiveness the water scarcely touched its original bed for many miles. Throughout the whole distance it was turned aside by each successive dam, carried along parallel to its natural course by canals and flumes, and then turned back to

[3] "River Mining," *Hutchings' Illustrated California Magazine*, II (1857-58), 98.
[4] *Ibid*.

the river bed so that it might be subjected to the same treatment by the next company.[5]

Obviously it would have been more efficient for the different companies on each river to have united their efforts so as to create a single and permanent drainage system, but there is no record of any such logical action being taken. One writer claimed that in a ten-mile section of the Feather River one dam was being built for every mile, at a cost of $8,000 apiece.[6]

The total amount spent on the various projects on each river sometimes was very large, especially after the size of the individual projects began to assume increasingly ambitious proportions in the middle fifties. In 1853 it was said that nearly twenty-five miles of the Yuba River had been turned aside, at a cost of $3,000,000, and in 1854 an Eastern mining consultant reported that over $1,500,000 had been spent for similar purposes on the American River and its branches. In 1855 seventeen fluming companies on the forks of the Feather River had a total capital stock of $303,000, and there were enough additional companies to increase the figure to $323,000.[7]

Sharp extremes of success and failure marked the fate of many of the river companies. Bad management and

[5] San Francisco *Daily Alta California*, July 17, 30, 1851.
[6] "River Mining," *Hutchings'*, II, 98-100.
[7] *Sacramento Weekly Union*, June 18, 1853; Josiah D. Whitney, *The Metallic Wealth of the United States, Described and Compared with that of Other Countries* (Philadelphia, 1854), p. 143; *Sacramento Daily Union*, July 23, 1855.

injudicious selection of sites barred many from success, while in the early years even wisely administered projects were made unnecessarily expensive by the use of canals rather than wooden flumes. Experience gradually demonstrated that despite the high cost of lumber it was cheaper to build flumes, "because the banks of the mining-streams are usually so steep, high, rocky, and crooked." [8] In order to construct a canal it was generally necessary to cut a part of the route through solid rock—a circumstance which made it essential to use gunpowder for blasting. This was apt to mean incurring a prohibitively heavy debt. By 1852 the greater cheapness of flumes had been established, and by 1854 the preference for them had become universal.[9]

Sometimes the cost of fluming was avoided entirely by building what was known as a "wing dam." This was a dam that was shaped like an "L." It extended from the shore out to the middle of the stream, and at that point it was turned at right angles and built down stream, thus leaving bare one-half of the river bed.[10]

Probably the most notable undertaking by a single company was the Cape Claim, on the Feather River. This was a great enterprise that was embarked upon at a comparatively late date—1857—and was thereby able to make maximum use of all the techniques that had been

[8] Hittell, *Resources of California*, p. 317.
[9] *Sacramento Weekly Union*, April 10, October 16, 23, 1852, November 15, 1856; *Sacramento Daily Union*, June 3, August 3, 1854.
[10] Woods, *Sixteen Months*, pp. 153-157.

developed during the previous eight seasons of river mining. It represented the culmination of the art of that branch of the gold industry. For it a construction contract of $120,000 was let in the latter part of 1856. The contract provided for the building of a wooden flume 3,800 feet long and 40 feet wide, and for the installation of eight sluices and more than a dozen water-powered pumps. The pumps were intended to free the claim of any water that was not drawn off by the diversion operation, as well as to ensure it against interference by the inevitable seepage through the dam.

When the water had been successfully deflected, 260 men were employed daily in digging "dirt" and washing it through the sluices. The daily operating cost during the period of actual mining was $1,500, but for thirty-five days the average yield per day was over $7,000. The total expenses were reported as $176,985.63, as against a total income of $251,725.45. But, as if to demonstrate the fickleness of river mining, the Cape Claim failed to meet expenses when the proprietors tried to repeat their performance in the following year.[11]

Contemporary newspaper reports indicate that river mining reached its peak in 1855 and 1856.[12] During

[11] *Sacramento Weekly Union*, November 15, 1856; *Sacramento Daily Union*, November 9, 14, 1857; "River Mining," *Hutchings'*, III (1858-59), 346-347; *New-York Daily Tribune*, December 27, 1858; Browne, *Report* (1867), p. 23.

[12] *Sacramento Daily Union*, October 4, 16, November 10, 1855; *Sacramento Weekly Union*, November 15, 1856; "Placer Mining," *California Mining Journal*, I, 52; and cf. Browne, *Report* (1867), p. 23.

those two seasons the natural conditions were unusually favorable, since the rains held off until December and since the seasonal amount of rain was less than in any similar period during the decade, save for the phenomenal drought of 1850-51.[13] At the same time, the process of trial and error had by then made the river miners experts at their calling, so that they were able to take full advantage of the chance to reap a rich harvest.

Thereafter the timing and quantity of the rainfall were less auspicious, while the yield from the overworked river beds was, inevitably, on the decline. By 1859 the white miners had abandoned a large part of the American River, the original home of river mining, to the Chinese, and by the close of 1863 the Asiatics had inherited the greater part of the river claims throughout the state. The fact that the whites no longer desired the claims for themselves is conclusive evidence of the declining profitableness of this once great type of mining.[14]

The chronological development of river mining included an initial boom in 1849 and 1850, then several bad years that temporarily robbed it of public confidence, and, finally, success in the middle fifties.[15] Quartz mining followed the same pattern. It began in 1849,

[13] McAdie, "Rainfall of California," *University of California Publications in Geography*, I, 212.

[14] *Mining and Scientific Press*, November 16, 1861, January 2, 1864; Browne, *Report* (1867), p. 23.

[15] *Sacramento Weekly Union*, November 15, 1856.

when gold in quartz veins, or lodes, was discovered at Mariposa. After a few months the natural public excitement over a new source of wealth was whipped into a speculative frenzy, and during 1850, 1851, and 1852 money and labor were poured into untried vein mining projects.

Not only in California, but also in the East and in Europe, impressive companies and corporations were formed by men who knew almost nothing about lode mining and still less about the location, potentialities, and peculiarities of California veins. Elaborate and costly machinery, often of untested design, was shipped to California, there to be set up and operated by amateurs.

The inevitable result was the collapse of the boom in 1852-53. The high favor in which quartz mining had formerly been held was converted into a wave of revulsion as strong as the hysteria that had preceded it. All but a few abandoned the new industry, and most of those who remained were practical mining men of comparatively small means.[16]

The survivors scrapped the impressive corporations and expensive machinery, reduced operations to a modest

[16] *Grass Valley Telegraph*, December 22, 1853, August 17, 1854, October 9, 1855, February 12, 1856; San Francisco *Daily Alta California*, August 25, October 28, 1852; *Sacramento Daily Union*, March 5, 1856, May 23, 1857; Ernest Seyd, *California and its Resources, A Work for the Merchant, the Capitalist, and the Emigrant* (London, 1858), pp. 44-49; Henry De Groot, "The Mother Lode of California," *Overland Monthly*, 1st series, IX (1872), 407-408.

scope, and set out to teach themselves in the fields of geology, mineralogy, engineering, and mechanics. The real beginning of lode mining in California is to be found not in 1849 and 1850, but rather in the half-dozen years that commenced in 1851-52 and extended throughout the ensuing period of depression and doubt.

Three separate operations are necessary in order to transform vein gold into marketable gold. The first is to "mine" the gold—that is, to burrow into the earth, and there to blast out or hew from the surrounding rock the gold-bearing parts of the quartz vein. The second is to pulverize these mined pieces of quartz, in order to break loose the gold particles. The third is to separate the gold particles from the other material, so that they may be "saved."

In California the first of the three did not raise problems as serious as one might expect. None of the California mines attained any great depth during the fifties, and for that reason they escaped the worst of the difficulties that develop when shafts are sunk far into the earth. Even after eight years of quartz mining, the deepest shafts in the state had penetrated to a depth of only a little over 300 feet, while in Europe one mine had been extended down to 3,778 feet, and many others to 1,500 feet. Indeed, many California operators worked their mines through a tunnel, or even through an open cut in a hillside, during part of the decade. Such conditions rendered relatively simple the usual problems of drainage, ventilation, hoisting, and timbering. Hand windlasses and moderate-sized steam pumps and hoisting engines seem to have been sufficient in most cases.[17]

In learning to operate underground, the Californians benefited greatly from the presence of trained workers from the coal, iron, tin, and lead mines of Europe and America. Cornish, English, and American veterans were on hand in considerable numbers at some of the most important quartz towns.[18]

When they found themselves facing the second of the three quartz operations—namely, pulverizing the rock —the Californians discovered that they could do best by focusing their efforts on improving certain ancient methods and machines whose basic principles had survived the hard test of centuries. Bitter experience during the early fifties revealed all too clearly the worthlessness of the ingenious new contrivances that were sent to California from Yankee and British workshops.[19]

The first and most important of the traditional machines was the stamp mill. This device came to California from the quartz mines of the southeast of the United States, but it had behind it a much longer history

[17] Cosmos (*pseud.*), "Quartz—Quartz Mining—Quartz Machinery," *California Mining Journal*, II (1857-58), 91; "Quartz Mining in California," *Hutchings' Illustrated California Magazine*, II (1857-58), 146-149; R. B. Noyes, "Early Mining and Milling Methods in California," *Mining and Scientific Press*, January 29, 1898.

[18] *Sacramento Weekly Union*, March 6, 1852; "Quartz Mining in Grass Valley," *California Mining Journal*, I (1856-57), 52.

[19] *Sacramento Weekly Union*, June 4, 1853; *Sacramento Daily Union*, May 23, 1857.

than the few decades during which it had been used in the South. As far back as the sixteenth century Agricola had given a clear, illustrated description of it, and in the years since then it had been used in many parts of Europe to crush not only gold quartz but also tin and lead ores.[20]

The stamp mill was not unlike a huge, mechanized version of a druggist's pestle and mortar. It consisted of a mortar, in which was placed the material to be crushed, and a heavily weighted pounder, called a "stamp." The "stamp" was a long, upright stem upon the lower end of which was placed a heavy iron head. The stamp rose and fell in response to the turning of a power-driven shaft to which it was geared by means of a cam. The cam kicked the stamp up to the top of its cycle, then released it so that it might fall with a crash upon the material in the mortar. The force of gravity brought the stamp's weight down with a pressure sufficiently great to cause solid rock to crumble into powder.[21]

Along with the stamp mill California also inherited from earlier mining countries a rotary contrivance that was intended to serve the same purpose. This had been developed in Spanish America, and was presumably the response to a need for a machine that was extremely

[20] Agricola, pp. 279-287; J. Arthur Phillips, *The Mining and Metallurgy of Gold and Silver* (London, 1867), p. 171; *Sacramento Transcript*, November 28, 1850.
[21] Description and diagrams in "Quartz Mining," *Hutchings'*, II, 145, 150.

simple in construction, that required a minimum of iron-work, and that could be operated by mule-power.

This machine was known as an *arrastre*.[22] To build it, flat-surfaced stones were fashioned into a circular track, around which a low retaining wall was built. In the center a strong post was sunk as a pivot, and to this a horizontal shaft was attached, like the arm on an old-fashioned marine windlass. Heavy abrasive stones were then placed in the track and connected to the shaft by long ropes or chains. A mule, plodding in a perpetual circle, provided the power to put the stones in motion, so that the gold-bearing material would be ground between them and the flat-surfaced bed.[23]

Sometimes this device was altered by substituting a heavy stone wheel for the arbasives. The wheel was pierced through the center like a millstone, so that the horizontal shaft could be thrust through the hole and the wheel thus be left free to revolve as the mule dragged it over the bits of auriferous rock. In such cases the contrivance was known not as an arrastre but as a "Chili mill." [24]

As between the stamp mill and the arrastre or Chili mill, the former was always the more popular with Americans. The arrastre was cheap and surprisingly effective, but it was too slow for the impatient "Brother

[22] Often Anglicized as "arastra" or "arrastra" by contemporaries.
[23] "Quartz Mining," *Hutchings'*, II, 151-153.
[24] *Ibid.*, p. 154.

135

Jonathan." Mexicans frequently did very well with it, and many Americans and Europeans used it temporarily, while testing a new vein or while laying by enough capital to buy a more pretentious outfit.[25] The Chili mill was regarded as being inferior to the arrastre in efficiency.[26]

During the height of the ill-fated boom, a group of inexperienced American gold seekers formed a company to open a quartz mine near the Cosumnes River. Their stamp mill was a crude affair that they had purchased from some Tennesseans. "It was just such a mill as had been made in Tennessee and other southern states previous to the discovery of gold in California," one of the group later recalled.[27] The mortar and the stems of the stamps were made of wood, and the shape of the stamps was square. The pounding heads of the stamps were made of soft iron.

When put into operation in the spring and summer of 1851, this primitive apparatus broke down so frequently that the members of the company found themselves spending half of each day in repairing it. Finally it occurred to one of them that a good deal of trouble could be avoided by substituting iron for wood in the construction of the stems, and since the corners of the heads

---

[25] Sacramento Weekly Union, June 4, 1853; Sacramento Daily Union, March 5, 1856; Phillips, Mining and Metallurgy, p. 170.

[26] Phillips, Mining and Metallurgy, p. 171.

[27] C. P. Stanford, "Origin of the California Stamp," Mining and Scientific Press, January 29, 1898.

were apt to be chipped off, it seemed reasonable to make both the stems and their heads round instead of square. A Yankee mechanic employed by the company made the further suggestion that the new round stamps would wear more evenly and operate more efficiently if allowed to revolve freely, instead of being held in place as had been customary.[28]

The resultant implement proved so successful that its use soon became universal. It has been known ever since as "the California stamp" and has been employed in all parts of the world.[29] It is one of the simplest yet most important contributions of California mining.

Other improvements soon followed, thanks to the operation of the process of trial and error. The soft iron heads of the stamps proved unsatisfactory wherever used in California, and were replaced at the earliest opportunity by heavy cast-iron "shoes." The shallow, unstable wooden mortar gave way to a deep, firmly fixed mortar that had a bedplate of cast-iron. Curved, tangential cams were substituted for the traditional straight ones. Patents were taken out on the basic device for an automatic ore feeder.[30]

[28] *Ibid.*

[29] Clarence King, "Introductory Remarks," Samuel F. Emmons and George F. Becker, "Statistics and Technology of the Precious Metals," *Tenth Census, 1880*, XIII, x; Knopf, "Mother Lode," U.S.G.S., *Professional Paper*, no. 157, p. 5.

[30] Stanford, "Origin of the California Stamp," *Mining and Scientific Press*, January 29, 1898; Almarin B. Paul, "Beginning of Quartz Mining in California: From 1848 to Discovery of the Comstock," *ibid.*; Noyes, "Early Mining," *ibid.* All three were pioneer quartz men.

As the result of these improvements and of others like them, the crude stamp mill of Agricola and the southern Appalachians was carried to a higher level of development during a half-dozen years in California than it had been during decades in its earlier homes.[31] Yet basically its principles were still those of the sixteenth century. A contemporary mining-town newspaper remarked:

All the great improvements in machinery have been sent to the foundry, to be re-cast into stamps [i.e., to be used as scrap iron], and the present mode of reducing the ore and saving the gold does not differ essentially from that practised at least two centuries back.[32]

Another mining-town journal provided the explanation when it said: "Men of experience have come to the conclusion that the simplest machinery is the best. Hence it is that stamps are now almost exclusively used." [33]

The factor of simplicity was especially important because of its bearing upon the purchase price of a quartz mill. During the period of speculation and elaborate contraptions, as much as $50,000 to $100,000 was often spent on a single plant, whereas in 1856 a completely equipped mill could be secured for $6,000 to $10,000.[34]

[31] Cf. Courtenay De Kalb, "The California Stamp-Mill," *Mining and Scientific Press*, May 21, 1910.
[32] *Sacramento Weekly Union*, January 24, 1857, clipping *Grass Valley Telegraph*, January 17.
[33] *Sacramento Daily Union*, March 5, 1856, clipping Nevada *Journal*.
[34] *Ibid.*; "Mining Summary—Quartz," *California Mining Journal*, I (1856-57), 4; cf. Cosmos, "Quartz," p. 91.

Part of this reduction was, of course, made possible by the general lowering of all charges in California after the flush years, but in any case the result was the same: by 1856 the necessary apparatus could be purchased without incurring a debt that would hound the miner into bankruptcy.

Progress in this phase of quartz operations was not accompanied by a similar advance in regard to separating and saving the gold from the pulverized rock. The latter was the most difficult operation in the whole process of lode mining, and for the improvement of it a greater amount of chemical and mineralogical knowledge was required than the average quartz operator possessed.

In the early years the methods used for saving the gold were almost unbelievably imperfect. Intelligent contemporaries estimated the proportion of gold saved at between 20 per cent and one-third of the total amount contained in the rock. Obviously, only the richest ores could be worked profitably upon so wasteful a basis.[35]

The enormity of this loss led to a constant searching for a remedy. A trial was made of all manner of devices, some novel, some as ancient as mining itself, but absolute success was not achieved within the decade. As had been the case with pulverizing machines, the

[35] Paul, "Beginning of Quartz Mining"; Jonas Winchester, Scrapbook of his correspondence for New York *Tribune*, p. 7, in California State Library, Sacramento.

operation of the trial and error process showed that, in general, the best results could be obtained by employing not innovations but rather some of the oldest methods and machines known to both lode and placer mining. The arrastre, for example, was found to be valuable as an adjunct to the stamp mill. Experience during the early fifties demonstrated that even the best stamp mills did not reduce the rock to a powder fine enough to free the maximum amount of gold. Accordingly, by 1856 it was becoming customary to subject the pulverized rock to a grinding in an arrastre after it had been taken from the stamp mill.[36]

Quicksilver, likewise, soon proved its worth as a means of salvaging gold from the pulpy mass which was formed when the ground-up rock was mixed with water. The employment of it became common at an early date. At some mills the quicksilver was placed inside the mortar, but at most it was used in conjunction with a variety of riffle boards, sluices, "amalgamating boxes," and "shaking tables" in a manner suggestive of placer mining.[37]

Still another useful device was to make the pulpy mass of pulverized rock and water flow over coarse blankets, since the interstices of the latter tended to catch gold

[36] "Quartz Mining," *California Mining Journal*, I, 20; "Quartz Mining," *Hutchings'*, II, 151-154.

[37] "Quartz Mining," *Hutchings'*, II, 150-151; "Quartz Mining," *California Mining Journal*, I, 20; *Sacramento Weekly Union*, June 11, 1853, February 26, 1859.

that was too fine to be retained by riffles.[38] This, too, was an ancient technique that antedated California by many centuries. It was probably a similar use of a sheepskin that gave rise to the myth of the Golden Fleece of antiquity.

Improvements such as these helped materially to reduce the ratio of loss, and the credit for adopting them must go to the experiments of the self-taught practical mining men. There was a point, however, beyond which such experiments became inadequate for the task at hand. A small percentage of the gold was enclosed within various forms of metallic sulphides colloquially known as "sulphurets." These had to be "broken down" if the gold within them was to be released, and in order to do so resort had to be made to methods borrowed from chemistry and mineralogy rather than from mining. The mechanical reduction of quartz rock was not sufficient, and mercury refused to act upon the gold in the presence of sulphides.[39]

During the early fifties, these intricacies were puzzling enough to discourage the average quartz operator from devoting much attention to the problem. The average operator of that period was preoccupied with the more pressing need for mastering the fundamentals of his trade. He was also buoyed up by the fortunate circum-

[38] "Quartz Mining," Hutchings', II, 150-151; "Quartz Mining," California Mining Journal, I, 20.
[39] Phillips, Mining and Metallurgy, pp. 190-198; Lindgren, "Gold-Quartz Veins," U.S.G.S., 17th Annual Report, p. 125.

stance that the ores near the surface had been sufficiently oxidized by weathering to be relatively free of sulphides. Only as depth was attained did the percentage of sulphides become large enough to arouse serious concern.[40] It may thus be taken as a sign of approaching maturity that in the last few years of the decade there was a marked rise of interest in eliminating this cause of loss.

The ultimate solution lay not in the use of mechanical contrivances but rather in the discovery of a chemical process, and no decisive progress was made until that was realized. A suitable process, not intended solely for gold, had been invented in Germany in 1848 and almost simultaneously in England. It involved heating the "sulphurets" in a furnace, then exposing them to penetration by chlorine gas, so as to convert them into a soluble chloride, and, finally, precipitating the resulting solution in order to extract the gold.[41] The introduction into California of this "chlorination process," as it was called by contemporaries, came through the medium of a group of skilled metallurgists and assayers who were certainly of German ancestry and probably of German training. These men put a small plant into operation in 1858 and labored diligently to adapt the process to California needs, but the complexity was so great that they were unable to perfect the method until after 1860.[42]

[40] Paul, "Beginning of Quartz Mining," in part.
[41] "Gold," *Encyclopaedia Britannica* (11th ed.), XII, 198.
[42] *Bean's History*, p. 127; Paul, "Beginning of Quartz Mining." The process is often called the Plattner process, after its German inventor.

Despite this delay in bettering the final step in quartz mining, that industry as a whole was in a strong position in the late fifties. At the start of the decade it had been a science whose mysteries were fully known to none in California and partially known to only a few. Now it was a tested business that could look for leadership to the graduates of its own school of hard experience. The most immediate of the technical problems that it presented had been overcome, thanks to the persistence and practical intelligence of a few hundred men. Wages and interest had declined, and the cost of all types of equipment and supplies had been greatly reduced. Untouched quartz veins were known to exist in many of the foothill districts, and the history of other parts of the world had shown that a good vein would last for a greater number of years than any other type of deposit.

The traditional forms of placer mining, by contrast, were on the decline in all save a few counties. The signs were plain that men must begin to look about for a new field for their energies. Quartz seemed an obvious possibility. The result was that from 1855 to the end of the decade the revival of interest in vein mining was the most widely discussed development in the California mineral industry.

The state geologist reported that there were only 39 quartz mills in successful operation in 1854, and according to the governor of the state the total fell to 32 before another twelve months had passed. Then the upswing began. In 1856 the number of mills rose to 59. In 1857 it was either 138 or 152, depending upon whose data you accepted. By November 1858 the total was 279, and by 1861 it was estimated at 300. These figures were for stamp mills. The count made in 1858 showed that there were also 519 arrastres, 310 of which were run as adjuncts to stamp mills.

The cost of erecting the 138 mills reported in 1857 was $1,763,000; the corresponding expense for the 279 mills of 1858 was $3,270,000. In both cases the total amount of capital invested in the industry must have been more than double the cost of the mills alone.[43]

These statistics disclose both the decided growth in the late fifties and the weak condition of the industry prior to that time. A comparison of these financial details with those previously given in connection with river mining should be enough to indicate the dominance of placer mining during the decade. By way of further evidence a report of 1856 indicates that less than one-twentieth of the state's working miners were engaged in quartz mining; and a report of 1857 gives the number as 8,300 or 8,400, less than a tenth of the total number

[43] Statistics from: *The State Register, and Year Book of Facts: for the Year 1857* (San Francisco, 1857), p. 215; the same for 1859 (San Francisco, 1859), p. 255; Hittell, *Mining in the Pacific States*, p. 38; "Mining Summary—Quartz," *California Mining Journal*, I, 4; "Quartz Miner's Convention," *ibid.*, I, 89; John B. Trask, "Report on the Geology of the Coast Mountains, and Part of the Sierra," California Senate, *Document*, 5 sess., 1854, no. 9, p. 90.

of working miners.[44] A statistician of the state mining bureau has even gone so far as to assert that from 1848 through 1850 100 per cent of the gold produced in California came from placer mines, and that from 1851 through 1860, 99 per cent came from that source.[45] Since so much of the time, money, and effort prior to 1860 was spent on speculation, experiment, and basic development, and so little on actual production, this calculation may not be as exaggerated as it might at first appear.

While these statistics reveal the rise and relative importance of the industry, it is well to remember that the whole story has not been told when one has stated the amount of gold produced or the number of men and dollars employed. That first ill-fated boom left almost no mark upon the statistical record, but it left a heavy impress upon people's lives. Perhaps nine out of every ten quartz schemes of that period ended in utter failure.[46] For many years thereafter the mining towns were veritable graveyards of costly, imported machinery that had been abandoned by despairing owners. Nor were the effects of the disaster confined to California. "In the

---

[44] *Sacramento Daily Union*, March 5, 1856; "Quartz Miner's Convention," *California Mining Journal*, I, 89.

[45] James M. Hill, "Historical Summary of Gold, Silver, Copper, Lead, and Zinc Produced in California, 1848 to 1926," U. S. Bureau of Mines, *Economic Paper*, no. 3 (Washington, 1929), p. 5, quoting estimates by Charles G. Yale.

[46] *Sacramento Weekly Union*, June 4, 1853.

beginning of 1853, there were at least twenty Anglo-Californian gold-quartz mining companies in the London market, representing nearly 2,000,000 shares, and the investment of about $10,000,000." [47] Few of these returned anything but disappointment to their British stockholders.

[47] Whitney, *Metallic Wealth*, p. 142.

X

TOWARDS A NEW ECONOMY: 1851-1860
II. DEEP DIGGINGS AND THE WATER SUPPLY

The delayed development that marked the history of quartz and river mining was duplicated in the evolution of the third major branch of the mineral industry. "Deep mining" meant the exploitation of buried deposits that had been laid down by rivers in earlier ages and subsequently covered over by debris. The Tertiary gravels were by far the most important of these.

Deep mining for the Tertiary gravels began in Nevada County in 1850, and it seems to have begun elsewhere at about the same time. Cornishmen are credited with suggesting the presence of the buried deposits and with making the first attempts to reach them. The Cornishmen, approaching the problem as veteran underground miners, sank short shafts into the hillsides until they struck "pay dirt." Popular usage promptly termed these small shafts "coyote holes." [1]

From this crude beginning deep mining, as practiced by means of shafts and tunnels, spread in a few years to many parts of the Sierra foothills and to the northwest. As was true of the other two major forms of mining, its history during the early fifties was marred by a multi-

[1] *Sacramento Transcript*, October 19, 1850; Borthwick, *Three Years*, p. 138; *Bean's History*, pp. 30, 65.

tude of mistakes, caused, in this case, largely by lack of experience and judgment in trying to guess the location of buried channels. An outcrop of gold on the face of a hill usually gave the miner his "lead," but in digging a tunnel to trace the lead back into the hill, the miner had to gamble his intelligence against the unpredictable whims of the lost river.

The financial losses during the early years were often very large, but they were insufficient to prevent "tunnel mining" from becoming by 1854 one of "the most important and productive of any branch of mining in the State." [2] By that time the customary California process of trial and error had taught many lessons, and by applying what they had learned, the miners of the middle and later fifties were able to undertake some highly impressive projects.

A report made in the early part of 1852 suggests that at that period few tunnels had been driven into the side of a hill for more than two hundred feet. Even with that shallow depth, however, the cost was five or six dollars per linear foot under good conditions, and as much as sixteen dollars per foot when tunneling through solid rock. [3]

By the latter part of the decade such figures had become picayune. From 1856 to 1860 one hears repeatedly of tunnels that were from one thousand to two

[2] Sacramento Daily Union, October 23, 1854.
[3] Sacramento Weekly Union, March 20, 1852.

thousand feet long, and of some that were nearer three thousand. It was said in 1861 that in a single district of three or four miles square there could often be found as many as fifty tunnels, each from three hundred to two thousand feet long, most of them cut part way through solid rock, and each costing from three to fifty dollars per linear foot, with the average at perhaps twenty dollars.[4] Since each had to be made big enough to permit the passage of the miners themselves and of the little railroad cars in which they hauled the excavated material, the work required was very large. In running a nine-hundred foot tunnel into the famous Table Mountain, with much of the way through solid rock, one company invested a collective total equal to 3,756 days of labor.[5] A mining-town newspaper said in 1856:

There are hundreds, and perhaps we might say thousands of tunnels, which have been from one to two years and more in progress, night and day, which have not yet reached the point sought for, but known to exist, rich in the precious dust. Money and labor enough has thus been invested by the miners of our State to build two such cities as San Francisco.[6]

Like the river-bed claims, the tunnel projects were usually undertaken by joint-stock associations composed principally of working miners. Since the members of the companies often had to labor for several years before

[4] *Mining and Scientific Press*, August 17, 1861.
[5] "Table Mountain from the Montezuma House," *Hutchings' Illustrated California Magazine*, I (1856-57), 544.
[6] *Grass Valley Telegraph*, February 12, 1856.

they struck "pay dirt," financial embarrassment was almost universal, and more than one company lost its tunnel to its creditors in the very moment of victory. Others had the equally hard luck to waste their capital and several years of effort on claims that proved to be barren. In the vicinity of Iowa Hill, Placer County, ninety tunnels had been attempted by July 1856. Of these, thirty-four had been abandoned as hopeless, twenty-three were still being continued, although without immediate reward, and thirty-two were returning some income.[7]

Such figures make it clear that tunneling was an extravagantly wasteful method of working the deep deposits. It survived, and even flourished, because the miner had to have some way of cutting through the many feet of debris that covered the treasure. With miners it was an axiom that the gold was always concentrated on the bed rock, and the researches of modern geologists have shown conclusively, in the case of the Tertiary gravels, that the lowest part of an auriferous hill is the richest. In a bank that might be from fifty to three hundred feet high, for example, the top gravels would yield between two cents and ten cents per cubic yard of "dirt," while the gravels

[7] Modern usage often terms this "drift" mining. Technically the main adit driven into the hill is a tunnel, and the subordinate passageways thrust out from it along the lead are drifts. On Iowa Hill, see *Sacramento Weekly Union*, July 12, 1856. On tunnel mining in general, see *ibid.*, July 5, August 2, 1856; *Sacramento Daily Union*, August 11, 1857; Grass Valley *Nevada National*, May 12, 1860; *Mining and Scientific Press*, August 17, 1861.

near the bed rock would yield from fifty cents to fifteen dollars for the same unit.[8]

This meant that if an auriferous bank were one hundred feet high, and a miner did not wish to resort to a tunnel, he would have to shovel away more than ninety feet of barren or low-grade material before he reached "pay dirt." The latter would have to be exceedingly rich to compensate for so much labor.

At first the gold seekers had no especial method for dealing with this problem other than burrowing into the earth. During 1850 some of them tried digging away the hillsides with pick and shovel and washing the debris in a long tom.[9] Early in the next year the improvement in mining implements came to their aid by producing the wooden sluice. With this new instrument a miner could wash approximately twice as many cubic yards of dirt in a day as with the long tom.[10] Even this was not sufficient, however, for it brought relief only at the final stage of the operation. The preliminary task of digging out the hillside and shoveling it into the sluice remained.

In order to lessen this preparatory labor, recourse was had to a practice that soon became known as "ground sluicing." To make use of this technique, the miner dug a small gully down the hillside that he intended to wash.

[8] Lindgren, "Tertiary Gravels," U.S.G.S., *Professional Paper*, no. 73, pp. 66, 71.

[9] Augustus J. Bowie, *A Practical Treatise on Hydraulic Mining in California. With Description of the Use and Construction of Ditches, Flumes, Wrought-Iron Pipes* (New York, 1885), p. 48.

[10] Browne, *Report* (1867), p. 22.

He then had a supply ditch or flume extended to the top of his hill, and presently he would send water cascading down the gully. Trusting to rocks and other obstructions to serve as natural riffles, the miner would then stand on the banks of his artificial watercourse and shovel and thrust masses of earth down into it. At intervals of a few weeks or months he would use a long tom or board sluice to "clean up" the fine debris that had accumulated behind the obstructions in the gully. This fine debris, of course, contained in concentrated form all the gold that had been saved.[11]

The ground sluice undoubtedly lost a good proportion of the gold that passed through it, but it was the best available means for handling the low-grade top gravels, and for that reason was used extensively during 1851 and 1852.[12] While employing it in Nevada County, in the spring of 1852, a Frenchman named Chabot decided to render his task somewhat easier by piping the water from the flume to his claim through a short length of hose. Because of its greater flexibility, this was more convenient than confining oneself exclusively to the use of ditches and flumes.

A year later Edward E. Matteson, a Connecticut Yankee who was working in the same area, had the simple but revolutionary idea of attaching a nozzle to the hose

[11] "Mining for Gold," *Hutchings'*, II, 8; Hittell, *Mining in the Pacific States*, pp. 141-142.
[12] San Francisco *Daily Alta California*, August 26, 1852; Rolfe, "Mines and Mining," *Bean's History*, p. 62.

and directing a powerful stream of water against the hill-side.[13] Therewith "hydraulic mining" came into being, and California made her greatest and most famous contribution to the world-wide science of extracting wealth from the earth.[14]

The action in "hydraulicking" was precisely what would be achieved if one directed a fire hose against a sand pile. It was so simple that one can hardly call it an "invention," and there is evidence to show that others, in earlier times, had given thought to the dynamic possibilities of a stream of water under pressure. Pliny, in his famous *Natural History*, says that the miners of his day sometimes created artificial waterfalls for the purpose of disintegrating auriferous material. Brazilians are believed to have practiced a somewhat similar method, and a southern historian has maintained that an unsuccessful trial of the process was made in North Carolina prior to the discovery of gold in California. Within California itself, two other persons have claimed the honor of having first devised and used a hydraulic hose. Others than Matteson, then, probably deserve a share of the credit, but it was Matteson who made the process known to the world, and it is Matteson whom Californians have always regarded as the father of hydraulicking.[15]

[13] *Ibid.* Contemporary spelling of his name varies.

[14] King, "Introductory Remarks," Emmons and Becker, "Statistics and Technology of the Precious Metals," *Tenth Census, 1880*, XIII, x.

[15] Pliny, *Natural History*, book xxxiii, chap. 21; Hittell, *Mining in the Pacific States*, p. 36; Green, "Gold Mining," *North Carolina Historical Review*, XIV, 150-151; Henry De Groot, "Hydraulic and Drift Mining,"

In the hydraulic the miner found precisely the instrument that he had been seeking. In the sluice he already possessed a method for washing large quantities of dirt at low cost. Now he acquired a cheap way of preparing dirt for the sluice. As a government report stated: "The man with the rocker might wash one cubic yard of earth in a day; with the tom he might average two yards; with the sluice four yards; and with the hydraulic and sluice together fifty or even a hundred yards." [16] The resultant saving in expense was very great. A mining engineer estimated that if wages were $4 a day, the cost of washing one cubic yard of auriferous earth with the pan would be $20; with the rocker, $5; with the long tom, $1; and with the hydraulic, $.20. [17]

The potentialities of the new process as a means of saving labor and time were recognized almost at once, and a contemporary description shows how effectively the hydraulic was being used at a few claims a year after it was discovered:

We stated yesterday that a claim at Iowa Hill had been worked into the hill by the application of hydraulic power, until it was *one hundred and twenty feet from the top of the*

California State Mineralogist, *Second Report, from December 1, 1880, to October 1, 1882,* appendix, p. 149.

[16] Browne, *Report* (1867), p. 22.

[17] W. S. Keyes, "Mineral Resources of the State of California," in *The Pacific Coast Business Directory for 1867: Containing the Name and Post Office Address of Each Merchant,* comp. by Henry G. Langley (San Francisco, 1867), p. 60.

Phillips, *Mining and Metallurgy,* p. 161, gives these same figures except that 5 cents is substituted for 20 cents.

hill to the bed rock in the claim. With a perpendicular column of water 120 feet high, in a strong hose, of which they work two, ten men who own the claim are enabled to run off hundreds of tons of dirt daily. So great is the force employed, that two men with the pipes, by directing streams of water against the base of the high bank, will cut it away to such an extent as to cause immense slides of earth, which often bring with them large trees and heavy boulders. To carry off these immense masses of dirt they have constructed two sluices, one for the paying and the other for the non-paying dirt. . . . After these immense masses of earth are undermined and brought down by the streams forced from the pipes, those same streams are turned upon the tons of fallen earth, and it melts away before them, and is carried away through the sluices with almost as much rapidity as if it were a bank of snow. No such labor-saving power has ever been introduced to assist the miner in his operations.[18]

Despite its acknowledged efficacy, the hydraulic process did not pass into general employment until several years had elapsed.[19] Its adoption by areas that had deep gravels was restricted by a number of limiting factors, one of which was the problem of securing an adequate supply of water. Hydraulicking required both a large volume and a large "head," or pressure, of water. The latter prerequisite automatically confined hydraulicking to hilly regions where a sufficient fall could be obtained.[20] The former could be provided only by building ditches, flumes, and reservoirs.

[18] *Sacramento Weekly Union,* July 22, 1854. Italics in text.
[19] *Grass Valley Telegraph,* January 31, April 4, 1857.
[20] *Sacramento Daily Union,* July 10, 1854.

Such structures could not be created overnight in response to the sudden appearance of hydraulicking. It is significant that both the origin and early development of the new form of mining centered in Nevada County, which was always the banner county in water companies. When a movement to make a more widespread use of the process developed in the latter part of 1856 and the early months of 1857, it was preceded by the construction of greater "facilities for saving and turning the water to useful account, by ditches and reservoirs," than had "existed at any other time" previously.[21]

Beyond the question of the water supply, there were further limitations that arose from the imperfectness of the hydraulic process during its early years. The most fundamental was in regard to the hydraulic unit itself. As first used, the unit was little more than a rawhide hose with a wooden nozzle. Canvas was soon substituted for the former and iron for the latter, but such simple changes were not sufficient to produce an implement that would survive the strain of operation under high pressure.

Since the weakest link in the system was the canvas hose, the obvious initial step was to reduce its length to a minimum. Attempts were made during the first season of hydraulicking to replace the greater part of the hose with iron pipe, and after several years of experimenting a type of pipe was developed that would withstand pres-

[21] *Grass Valley Telegraph*, April 4, 1857.

sure and rust and yet would not be prohibitively expensive.[22]

The hose could not be dispensed with entirely since there had to be some way of aiming the stream of water. The miners reduced the canvas to a short length, and for this they contrived something that was appropriately nicknamed "crinoline hose." This was canvas hose that had been strengthened by a succession of encircling iron rings, or by rope bindings or strong netting.[23]

The other improvements in the hydraulic process were concerned not with the hydraulic unit itself but rather with the more efficient treatment of the gravels against which hydraulicking was used. In particular, it was necessary to devise ways of dealing with a resistant, finely compacted type of gravel that was colloquially known as "cement."

When whole banks of this tough substance had to be attacked, the rate of speed of hydraulicking was slowed down so much that the process became almost unprofitable. Ideally, the miner liked to have a hillside crumble so quickly that the same stream of water that disintegrated it would sweep the auriferous gravel through the sluices. Since water often cost a company $45 a day, it was important to keep the sluices running without interruption.

[22] "Galvanized Iron, and Common Sheet Iron for Hydraulic Purposes," *California Mining Journal*, I (1857-58), 92; Bowie, *Practical Treatise*, p. 49; Wilson, *Hydraulic and Placer Mining*, pp. 52-53.
[23] *Mining and Scientific Press*, September 21, 1863; Phillips, *Mining and Metallurgy*, pp. 155-156.

When "piping" against a hard bank, however, the material was likely to yield so slowly that the sluices would be idle for one-half to three-quarters of the time.[24]

A satisfactory way of dealing with this seems not to have been contrived until 1858. In that year the miners found that they could greatly expedite matters by driving a small tunnel into the bank, and there setting off a charge of gunpowder. The blast left the hillside in such a loosened condition that it crumbled easily before an attack with the hydraulic hose.[25]

A second difficulty was caused by the tendency of the cement to break off in hard chunks within which the gold was securely locked. Although this complication was encountered as early as 1853, it was not resolved until 1857 and 1858. During those two years it was discovered that the chunks could be broken up by the use of stamp mills similar to those employed with quartz.[26]

A third point at which improvement was needed, both in dealing with cement and with less resistant forms of gravel, was in the final process of saving the gold from the disintegrated material. The wastefulness in this last step in hydraulicking was as great as in the corresponding operation in the quartz industry.

Even after half a dozen years of experiment, the

[24] North San Juan *Hydraulic Press*, October 9, 30, November 6, 1858.
[25] *Ibid.*, October 9, 30, November 6, 27, December 18, 1858, January 15, 22, 1859.
[26] Rolfe, "Mines and Mining," pp. 57-61; Grass Valley *Daily National*, January 4, 1865.

miners of the leading hydraulic town in the state were forced to admit that they were losing between a quarter and a half of the gold contained in each cubic yard of "dirt" washed, and a magazine editor claimed that over the state as a whole three-fifths of the gold was being lost.[27]

In their attempts to reduce the percentage of waste, the miners extended the length of their lines of continuous sluices to many hundreds of feet, in order to give the gold a greater opportunity to break loose from the lighter material and settle to the bottom, where it would be caught by the riffles and quicksilver. At first the lengthened sluices proved expensive, because the hydraulic stream swept into them great boulders that tore up the riffles and wore out the bottoms. By sheer accident it was found that this could be avoided by dispensing with the riffles entirely and, instead, lining the sluices with loose planks and wooden and granite blocks that would serve simultaneously as gold retainers and as protectors for the bottom of the sluices.[28]

Despite the greater length of the sluice lines, the loss continued. Still the mound of refuse, or "tailings," at the end of each line was enriched daily with gold that had been carried through by the high speed of the rush of rocks and water. While surveying the situation, it

[27] North San Juan *Hydraulic Press*, June 25, 1859; "Editor's Table," *Hutchings' Illustrated California Magazine*, III (1858-59), 432.
[28] *Sacramento Daily Union*, July 11, 1854; North San Juan *Hydraulic Press*, November 6, 1858.

occurred to an ingenious Nevada County miner that what was needed was to subject the auriferous material to sluicing under conditions that were less suggestive of a spring flood. To meet the need he designed the "under-current sluice." This involved placing a grating in the bottom of the regular sluice and drawing off through it the gold and other fine debris so that they could be treated in a second sluice underneath, where the use of a quieter current was possible. The large rocks and most of the water, meanwhile, boomed on down the regular sluice.[29]

This helped greatly, but it was still not a complete solution. The tailings continued to receive a goodly proportion of the miners' gold. Realizing that, alert mining men originated a new form of activity that was generally known as "tail sluicing." Special companies were formed for the sole purpose of building extensive lines of sluices—often several thousand feet long—through which to rewash the tailings. It was not uncommon for these companies to obtain a larger yield per ton than the hydraulickers who were dealing with the original deposits.[30]

The development of hydraulicking was intimately connected with the progress in extending the operations of the water companies. The earliest of the latter were

[29] Hittell, *Mining in the Pacific States*, pp. 143-144; Wilson, *Hydraulic and Placer Mining*, pp. 35-36, 98.
[30] North San Juan *Hydraulic Press*, November 6, 1858.

projected in 1850, and by the time that the hydraulic process was discovered some parts of the mines were beginning to be fairly well supplied, if judged by the needs of the long tom and sluice. The greatest number of the water companies were located in the central mining section. There the counties of Nevada, El Dorado, Placer, Yuba, and the upper portion of Sacramento all had several good-sized ditches that were in operation or under construction.

The northwest, at that period, was hardly in a position to support large enterprises, while in the Southern Mines there were only a few areas in which large or numerous undertakings were being attempted. The chief foci of southern activity seem to have been in Tuolumne County, where interest centered in the rich Columbia basin and the territory tributary to Columbia's neighbor, Sonora, and in Calaveras County, where Mokelumne Hill's lucrative diggings gave rise to a very ambitious project.

Most of the canals and flumes of 1850 and 1851 were small affairs, designed to convey water for only a few miles and to only a limited area.[31] During the next two years, however, projects were begun that by their scope would excite the interest of a modern engineer, while by their cost they would arouse the wonderment of a financier.

In El Dorado County, for example, the South Fork

[31] *Sacramento Daily Union*, December 14, 1857.

Canal was 16¼ miles long and cost $275,000. In Calaveras County the Mokelumne Canal Company's line of flumes and ditches was 18 miles long and cost $250,000. The Tuolumne Water Company, formed for the purpose of bringing water to Columbia from a point 20 miles away, spent $200,000 without achieving its aim.[32]

These were the largest rather than the average of ditch companies in the state at that time. Their excessive cost was in part a reflection of the prevailing high price of labor and supplies, in part an indication of the difficulty of the tasks undertaken, and in part the result of inexperience, undue optimism, and impatience with all economies that required a sacrifice of speed of construction.[33]

Over much of their length these projects made use of canals and ditches that were dug by the pick and shovel labor of gangs of workers. Where ravines or canyons crossed the route, expensive wooden flumes or aqueducts were built. If a ridge was in the way, it was tunneled through.[34]

The men who planned these great ventures had reason to expect a ready sale for all the water they could provide, since the experience of the pioneer water companies had shown that the demands of the miners' "toms" and sluices were insatiable. With the advent of hydrau-

[32] *Sacramento Weekly Union*, May 1, August 21, September 26, December 11, 1852, June 11, August 6, 1853.

[33] *Ibid.*, October 30, 1858; *Sacramento Daily Union*, May 4, 1854, October 25, 1855; *Grass Valley Telegraph*, March 18, 1856.

[34] *Sacramento Weekly Union*, January 24, February 28, May 1, October 16, 1852, August 6, 1853.

licking, the need for water increased fifty-fold,[35] but by a stroke of ill-fortune the discovery of the hydraulic process happened to coincide with the end of one of the wettest seasons in California's history and with the beginning of a dry cycle that produced a smaller amount of rain in each of four successive seasons.[36] The concurrence between the enlarged demand and the diminishing rainfall produced a continual cry of drought that sometimes was loud enough to blind contemporaries to the really astonishing progress of the water companies during the middle fifties. As the Nevada *Journal* pointed out in 1857:

The present mode of mining is so very different from the operations of a few years ago, that although the supply of water doubles every year, from the increased number and capacity of ditches, still the deficiency seems to be as great now as ever. The amount of water now used by one company of four or five men, two or three years ago would have kept two or three hundred men at work. The main difference is that *then*, men did the work, using the water merely as an auxiliary, and the amount of work done depended on the number of men employed. But now, the water is used as the laboring agent, and like machinery in manufacturing, only men enough to keep the machinery properly directed, are required. Thus banks of earth that would have kept a hundred men employed for months in its removal, will now be removed by three or four men in two weeks.[37]

[35] *Ibid.*, March 28, 1857.
[36] McAdie, "Rainfall of California," *University of California Publications in Geography*, I, 212.
[37] *Sacramento Weekly Union*, March 28, 1857, clipping Nevada *Journal*, March 20.

The truth was that the ditch companies had thrust forward their water systems so rapidly that in 1857 there were 4,405 miles of canals, ditches, and flumes, constructed at a cost of $11,890,800, and in 1859 there were 5,726 miles, built by the expenditure of $13,575,400. At the latter date there were also one thousand miles of small branch lines connecting with the main ditches. Furthermore, the size of the largest individual projects during the later years was such as to dwarf the corresponding enterprises of 1852 and 1853.[38]

The biggest in the state in 1856 was the Eureka Canal, in El Dorado County. Its main canals and branches extended for 247 miles, and they, together with the accompanying dams and reservoirs, had cost $700,000. Another huge company was one begun in Nevada County in 1855 by a group of courageous Frenchmen. Before its completion in 1860, this water system was to cost $950,000. Then there was the case of Tuolumne County, where $1,000,000 and four years of labor were spent to bring water to Columbia.[39]

Along with the increased size of the later projects went a greater boldness of conception and a higher degree of engineering ability, thanks partly to experience

[38] *The State Register . . . for 1857*, p. 230; *The State Register . . . for 1859*, p. 275.
[39] *Sacramento Weekly Union*, May 3, 1856; Rolfe, "Mines and Mining," p. 66; [Herbert O. Lang], *A History of Tuolumne County, California. Compiled from the Most Authentic Records* (San Francisco, 1882), p. 177.

and partly to the advent of trained engineers. Aqueducts, for example, became highly impressive affairs. In order to carry the water from the summit of one hill to the summit of another, a suspension flume was built in Yuba County that was 1,500 feet long and 206 feet above the ground. It rested upon the tops of the trees as it spanned the intervening valley. When a deep chasm had to be crossed along the course of the Feather River, the ingenious builders ordered from San Francisco 8,000 feet of boiler-iron pipe and fashioned it into a huge inverted siphon.[40]

The ambitiousness of these projects often proved their undoing. All too frequently the planners of a water system underestimated their costs. They would spend all of their money on the first three-quarters of the construction work, and then would have to borrow to complete the remainder. With interest at 3 per cent a month, it required only some unforeseen delay to bankrupt the promoters and either give ownership to the creditors, or else force the whole enterprise into a sheriff's sale. In this fashion the proprietorship of many, perhaps of most, of the water companies of the fifties passed from the hands of their originators.[41]

[40] "Suspension Flume across Brandy Gulch," *Hutchings' Illustrated California Magazine*, II (1857-58), 105; "The Frenchtown Canal Company of Butte County, Novel Ditch Enterprise," *California Mining Journal*, II (1857-58), 44.
[41] *Sacramento Daily Union*, May 4, 1854, December 14, 1857; *Sacramento Weekly Union*, January 28, 1854, October 30, 1858; *Grass Valley*

Since the latter were usually miners, engineers, and local merchants, while the new owners were what contemporaries called "capitalists," the operation of this process sometimes meant a transfer of control from the working men in the foothills to the business and financial men in the cities. The shift was, however, a gradual one and hardly reached significant proportions until the latter part of the fifties. Certainly in the middle of the decade "the greater part of the immense amount" invested in the water companies still belonged "to parties in the mines." [42]

That there was always some investment in water companies by men outside the mines is revealed by the personnel of two conventions that met in Sacramento. Both of these were composed of delegates from the various water companies. The first assembled in 1853 and the second in 1857. Present at the first were a prominent banker, who represented three different companies, a political officeholder, who also represented three, a San Francisco attorney, and two importers and wholesale merchants. At the second and more important of the two conventions, there were three judges, three bankers, three present or former political officeholders, two

Telegraph, March 18, 1856; "Suspension Flume across the Stanislaus River," *Hutchings' Illustrated California Magazine*, V (1860-61), 296; "Suspension Flume across Brandy Gulch," *ibid.*, II, 105; Rolfe, "Mines and Mining," pp. 66-67.
[42] *Sacramento Daily Union*, October 25, 1855.

transportation company officials, and one man who seems to have been a flour mill proprietor. There were also two individuals who apparently made a profession out of serving as city agent for mining concerns.[43]

While some of these men doubtless acquired their interest by buying out bankrupt companies, others must have been original investors. For example, the famous Sacramento bank of D. O. Mills & Co., which was represented at both conventions, was instrumental in enabling the miners of Columbia, Tuolumne County, to complete the first ditch into that rich basin. When the ditch project was about to fail for lack of funds, the bank bought a sufficiently large block of stock in it to make possible the consummation of the work.[44]

Similarly, on two different occasions water companies in the region tributary to Marysville sent agents down to their queen city to ask the business and financial men to subscribe to the company stock.[45] That such appeals for aid did not always go unanswered is to be inferred from the character of the board of directors of another water company in the Marysville area. All five of the directors

[43] Conclusions reached after checking the convention rosters, as printed in *Sacramento Weekly Union*, October 8, 1853, and *Sacramento Daily Union*, August 29, 1857, against the contemporary city directories of San Francisco, Sacramento, and Marysville.

[44] Heckendorn and Wilson, *Miners & Business Men's Directory*, p. 7; Thomas Conlin, "The Story of Columbia," *Pacific Underwriter and Banker*, XXXIX (1925), 334.

[45] *Marysville Daily Herald*, February 1, 1854; Marysville *Tri-Weekly California Express*, March 28, 1855.

of this concern were Marysville businessmen and bankers.[46]

This tendency for urban capital to find its way into the foothills via the water companies was at variance with the general trend. At the start of the fifties city merchants and financiers, especially in San Francisco, were badly "burned" when they advanced funds to unsound mining ventures, particularly to quartz companies. The lesson so learned was not soon forgotten. It had the result that bitterness, distrust, and mutual recriminations between the miners of the interior and the "capitalists" of San Francisco were thereafter a standard factor in the social and political life of the state.[47] On more than one occasion the urban "monied men" were caustically denounced for their alleged unwillingness to let their surplus resources seek investment beyond the corporate limits of the cities.[48]

This does not mean that the miners were left entirely upon their own. In many districts small sums were advanced by the local "tradesman, the mechanic, the lawyer, the doctor, and in some places even by the parson." [49] More important was the universal custom by

[46] Comparing board of directors, as printed in *Marysville Herald*, July 21, 1855, with Marysville city directories.

[47] Henry De Groot, "Mining on the Pacific Coast, Its Dead-Work and Dark Phases," *Overland Monthly*, 1st series, VII (1871), 152.

[48] *Sacramento Weekly Union*, January 21, 1854, February 17, 1855; "Capital in California," *Hutchings' Illustrated California Magazine*, I (1856-57), 130-132.

[49] *Sacramento Weekly Union*, August 23, 1856.

which the local merchants in the camps and towns gave credit for several months at a time, while they, in turn, secured credit from the city wholesalers and importers.[50]

By this means the miners received the use of a not insignificant supply of cash and credit with which to bolster their own earnings. In general, however, direct city interest in the mines after the early years was limited in extent and tended to be restricted to the purchasing of shares in water companies. Over the field of mining as a whole, and throughout the greater part of the transitional years from 1851-1860, the financial burden was borne by the miners themselves and by the local businessmen who sold them supplies in return for a promise to pay.

A recognition of this fact is necessary if one is to appreciate fully the significance of the miners' achievement during this decade. Within that brief period, despite a perennial shortage of cheap capital and labor, despite also the costliness of supplies and transportation, the miners brought into being two complex but successful forms of California mining: deep diggings and quartz.

For the former, insofar as hydraulicking was concerned, they created a process that was new to the world. For the latter they adapted and improved the practices of

[50] *Sacramento Daily Union*, January 26, 1856; *Sacramento Weekly Union*, December 18, 1858; "Placer Mining Summary," *California Mining Journal*, I, 21.

older countries. In the meantime they stripped the shallow placers of whatever gold the forty-niners had left, and they exploited the river beds in imperial fashion, only to see both types of mining decline into insignificance. Through all these changes the people of the foothills safely steered the industry upon whose continued prosperity depended the welfare of the whole state.

XI

GREATER CALIFORNIA

The history of each of the several branches of mining during the fifties was a demonstration of the accuracy of the *Alta California's* prediction of 1851 that "to get the gold from" the deposits which remained "we must employ gold." [1] The day had passed in which a man could win a fortune by the employment of no capital save the labor of his two hands. At the close of the season of 1856 the *Sacramento Union* observed: "As a general rule, but few places can now be found in the mines where, without the expenditure of considerable capital, individuals, or even companies, can make three dollars a day." [2]

For the individual miner this coming of the lean years meant not only the disappearance of the vision of wealth; it implied also the establishment of a modern capitalistic economy, in which a few owned the means of production and were thereby enabled to hire the services of the many. This John S. Hittell recognized when he wrote in 1858:

The business of mining in California has been declining constantly since 1851—at least as a source of profit for most of the

[1] See above, p. 117.
[2] *Sacramento Weekly Union*, September 6, 1856.

miners. . . . The main profits now go into the hands of a few who are in possession of rich claims, whereas in 1851 the profits were much more equally divided. Then no capital was required by the miner, and little experience; the best diggings were in the easily-obtained gravel in the beds of brooks and ravines, and on the bars of creeks and rivers. . . . Our gold comes now from deep down in the bowels of the earth, from tunnels, quartz veins, shafts and hydraulic claims, and when found near the surface, a large proportion of it must go to pay ditch companies for the water used in washing. In 1851 labor pocketed all the profits of the mines; in 1858 capital pockets most of it.[3]

The *Sacramento Union* gave a similar description of the change, and in the same year:

A complete revolution [has been] produced in the methods and means applied to mining. Formerly the gold was taken out by miners working independently and individually; but since the river bars and placer diggings were mostly worked over, claims—deep, quartz, river, tunnels, etc.—have gradually concentrated in the hands of men of means, who have employed others to mine for them.[4]

In short, as the San Francisco *Morning Call* expressed it: "The mines have ceased to be the poor man's friend." [5] Henceforth the impecunious old-timer and the penniless newcomer must either labor for hire or

[3] John S. Hittell's correspondence to the *New-York Daily Tribune,* December 27, 1858. Cf. editorial in *California Mining Journal,* I (1856-57), 12.

[4] *Sacramento Daily Union,* June 11, 1858. Cf. *Mining and Scientific Press,* June 1, 1861.

[5] *New-York Daily Tribune,* March 14, 1859, clipping San Francisco *Morning Call.*

else devote their efforts to that hardest and most ill-paid of occupations: "prospecting" for new mineral ground.[6]

Some were too restless and independent to accept the first of the two alternatives. The others resigned themselves to a future as someone else's employees. If thereby they lost their freedom, they gained much in return. To them the change brought greater certainty of income, greater fixity of abode, and hence greater regularity of life. For them, better built and more comfortable houses became possible.

With a more settled population as a market, local agriculture was encouraged to provide fresh vegetables, fruit, dairy products, and the like. Towns became something more than supply depots and centers for Sunday Saturnalias. The sight of women and children ceased to be a cause for delighted amazement. Communications with the outer world of the cities took on greater certainty and lost much of their old laborious slowness.[7]

Even if a man were willing to accept these gains as sufficient recompense for his lost freedom, the opportunity to make the choice was not always open to him. The achievement of greater stability was by no means universal in the mineral region. By 1858 the Southern Mines as a whole were entering upon a permanent decline, while in the central section of the Northern Mines

---

[6] *Stockton Daily Argus*, March 10, 1857.
[7] "Placer Mining Summary," *California Mining Journal*, I, 4, 13, 21; editorial in *ibid.*, p. 12; *Sacramento Weekly Union*, September 6, 1856.

many of the most famous districts experienced their last days of greatness before the fifties came to a close.

The new forms of mining—quartz, hydraulic, and tunnel—were highly selective in their influence. Quartz, for example, made Grass Valley, Nevada County, the most prosperous and most permanent gold town in the state, but it did little for Grass Valley's neighbors, Nevada City and Rough and Ready. Quartz also began to infuse a new life into some of the towns of Amador and El Dorado counties at the very moment in which the wealth of their placers was failing, but it did nothing to help communities adjacent to them. Hydraulicking brought into being in the central section, and especially in Nevada County, a succession of towns that hardly had an existence prior to 1853, but it was powerless to aid the older districts that did not possess the requisite natural conditions.

Each locality to which the late fifties brought decadence, either sent forth its discouraged miners to try their luck elsewhere, or else retained them, temporarily, in a state of discontent and penury. If a miner did seek employment elsewhere, his chances of success were diminished by the fact that the more favored districts were suffering from what modern economists would call "technological unemployment." As the *California Mining Journal* explained it:

There now is, and for some two years past has been much complaint among a large class of our population about a lack of employment; and this, among uninformed persons, has generally been attributed to a giving out of the mines. . . . [Actually] by the recent improvements which have been introduced into mining operations, a large amount of manual labor has been dispensed with. Four or five men will now do more work with a hydraulic hose than fifty men could do formerly with the appliances then in vogue. . . . [Accordingly,] the hydraulic hose is throwing multitudes of miners out of employment.[8]

By 1858 there were, therefore, many miners who were in need of work. Had they been willing to do so, these men could have found employment at low wages in the marginal placer diggings and in the quartz mines, or they could have changed their occupation to farming, lumbering, or one of the trades. California was still capable of absorbing a large amount of labor, provided always that it could be secured at comparatively low rates and in hired status.[9]

Such a means of relief had little appeal for the footloose miners. The latter wanted a continuance of "the wages to which they had become accustomed," [10] and they were psychologically unprepared for a return to the humdrum existence from which they had escaped when they left their eastern or European homes to seek the Golden Fleece.

[8] "Placer Mining Summary," *California Mining Journal,* I, 4.

[9] *Napa County Reporter,* January 23, 1858, July 2, 1859; *Sacramento Weekly Union,* December 4, 1858; Henry S. Brooks, "A Few Words about 'Pile' Making," *California Mountaineer,* I (1861), 130-131.

[10] John S. Hittell, "The Mining Excitements of California," *Overland Monthly,* 1st series, II (1869), 415.

They had become industrially desperate. They were ready to go anywhere if there was a reasonable hope of rich diggings, rather than submit to live without the high pay and excitement which they had enjoyed for years in the Sacramento placers. Many of them had become unfit for the placid and orderly routine of the common laborer in other countries.[11]

What these men wanted was a new El Dorado, a new golden land such as California had been in '48 and '49. This was no new desire. At intervals throughout the fifties the covert whisper of "gold! gold!" had produced ill-justified "rushes" to remote parts of California, to Oregon and Washington, and to South America and Australia.[12] The greatest of them, prior to 1858, had been the Kern River humbug of 1855.

Kern River was an almost unknown district that lay far to the south of the most distant boundaries of the Southern Mines. It possessed some gold, but not very much, and was beyond the reach of adequate means of transportation and communication. At the turn of the year 1854-55 a series of deliberately false letters to the San Francisco newspapers suddenly blared forth the "news" that fabulous wealth had been found on Kern River.

Five thousand men who were not even sure of the exact location of that stream promptly hurried off to win an easy fortune, and nearly as many more prepared to

[11] *Ibid.*
[12] A list of thirty "rushes" and "excitements," large and small, 1849-1862, is given in *Mining and Scientific Press*, July 16, 1862.

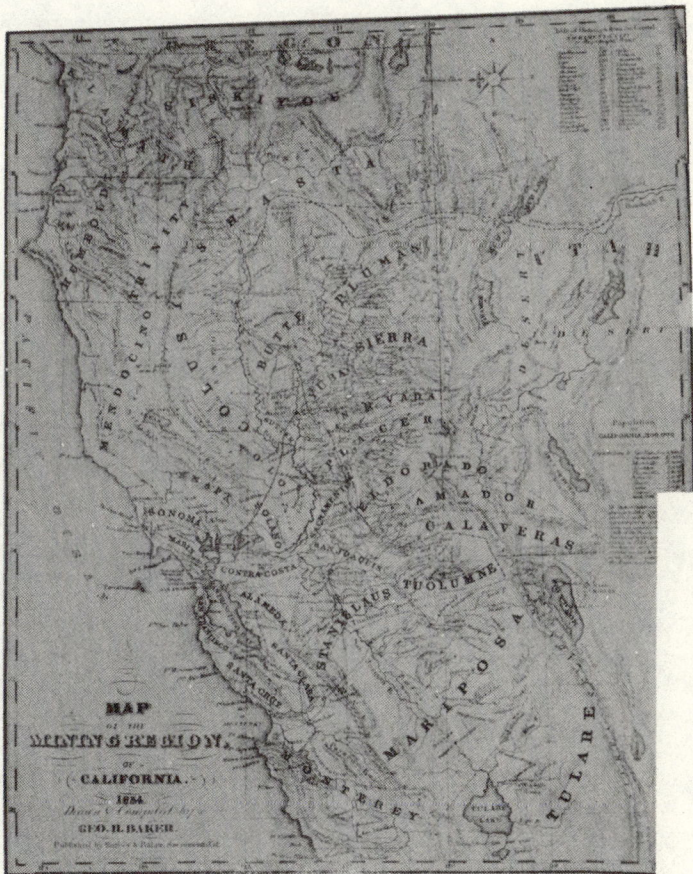

MAP
OF THE
MINING REGION,
OF
CALIFORNIA.
1854
Drawn & Compiled by
GEO. H. BAKER.

follow. Before the latter could get under way, bona fide reports came back from the first arrivals that the richness of Kern River had been grossly exaggerated and that the hardship of reaching it and maintaining oneself there had been greatly understated. The boom quickly collapsed, but before it did so, paying mines and farms had been vacated by men who were more willing to be ruled by optimism than by good judgment.[13]

If so great a rush could spring from such slight evidence in 1855, then it is not to be wondered that a similar rumor could create a veritable stampede three years later, when the number of unemployed, dissatisfied miners was much larger than it had been in 1855. This time it was Fraser River, in British Columbia, that was the new El Dorado. Psychologically Fraser River had the advantage over Kern River, for it was even further off and even less was known about its true richness or lack of richness.

Like all mass migrations, the Fraser River rush left behind it no precise statistics by which one can measure its dimensions.

The custom-house records say that between the twentieth of April and the ninth of August, the limits of the Fraser fever, fifteen thousand and eighty-eight passengers left San Francisco in one hundred and twelve vessels for the new Eldorado; but the 'Prices Current,' a carefully edited commercial journal, said the number of adventurers was twenty-three thousand

[13] On Kern River, see Hittell, "Mining Excitements," p. 414; *Sacramento Daily Union*, February 22, 24, 27, March 1, 2, 5, 6, 1855.

four hundred and twenty-eight, the reports to the custom-house being greatly below the truth in many cases.[14]

If the latter figure is correct, then six persons out of every hundred in the state must have left California within four months. It is understandable why it seemed to some that California was "in danger of being depopulated." [15]

Although city folk and farmers joined in the stampede, it was the mining counties that felt most severely the effect of the sudden exodus. From Northern and Southern Mines alike came reports of daily departures. A check of travel through Sacramento showed that during the thirty days which began on the twentieth of May, 3,669 more persons arrived from the Northern and Southern Mines, en route for San Francisco, than was usual during such a period. The new Argonauts took with them all the capital they could command, and behind them left declining property values.[16]

Fraser River, like Kern, proved to be 10 per cent truth and 90 per cent humbug. The significance of it lay in the drain which it caused upon the vitality of the bona fide mining regions, and in the new evidence which it gave as to the volatile nature of the California mining population.

[14] Hittell, *History of San Francisco*, p. 275.
[15] *Ibid.* The estimate of six out of one hundred is Hittell's.
[16] See *Sacramento Daily Union*, May 19, 29, June 3, 9, 18, 1858; *Grass Valley Telegraph*, June 12, 1858; Edwin G. Waite, "Historical Sketch of Nevada County, California," in *Bean's History*, p. 14.

Kern and Fraser had turned out to be false leads. What would happen if a genuine bonanza were discovered, especially if it were nearer home?

On June 24, 1859, a former Californian rode into Nevada City and called at the office of the Nevada *Journal*. He had just completed a one-hundred mile journey across the Sierras from what is now known as the State of Nevada. Over on that far side of the mountains a small band of gold miners had been carrying on limited operations for ten years, without achieving more than moderate success. Of late they had been annoyed by the intrusion into their gold leads of a bluish rock of a type that they had never before seen.

Curiosity finally prompted the dispatching to California of a bag of specimens of the troublesome "blue stuff." At the *Journal* office the messenger stopped long enough to tell the editor about the miners' problem, then he delivered the specimens for analysis by assayers in Nevada City and its neighbor, Grass Valley. On July 1 the *Journal* published the results of the assays: the "blue stuff" was silver, and the specimens were rich in it.[17]

Although Nevada County was the most prosperous gold county in California, many of its citizens at once prepared to leave for "Silverado," as they nicknamed the

---

[17] *Ibid.*, p. 14. Waite was the editor of the *Journal* at the time. The ill-fated Grosch, or Grosh, brothers had guessed the secret of the blue rock three years earlier but died without revealing their knowledge.

new Comstock Lode. Two parties, one from Grass Valley and another from Nevada City, started across the mountains immediately. "In a few days others were on the route; more soon followed, and within two years, it is probable one-third of the male adults of Nevada County had gone to the silver region, either to try their fortunes or visit the scenes that had created so intense an excitement." [18]

Elsewhere in California the news produced similar results. Before the November snows closed the Sierra passes, every restless miner who could walk or could secure transportation went hurrying over the mountains in true '49 style. During the ensuing months of winter the excitement swelled into a hysteria such as had not been seen since Marshall's discovery at Coloma. Bankers, merchants, clerks, and farmers joined in the stampede, and long before the warm spring sun of 1860 had melted the mountain snow, hundreds were trying to fight their way across the steep passes.[19]

These early sufferers from the "silver mania" were but the vanguard of an army that continued for three and one-half years to march eastward across the Sierras. Not until the latter half of 1863 did the newspapers report any significant reversal in the movement of population,

[18] *Ibid.*, p. 15.
[19] North San Juan *Hydraulic Press*, July 23, 1859, March 31, 1860; Eliot Lord, *Comstock Mining and Miners*, U. S. Geological Survey, Monographs, IV (Washington, 1883), pp. 57, 63-71, 96; Grass Valley *Nevada National*, February 25, 1860.

and not until a year later was there much indication of a permanent change in California's favor.[20]

In the meantime all parts of the California mining region suffered a never-ending drain upon their man-power and capital. Even camps that had paying diggings saw their restless citizens start on the eastward trek, while districts that were already on the downward grade were almost depopulated. Many camps never recovered; others regained but a portion of their former strength.

It was not only California men that crossed the mountains. California dollars were also sent to Nevada. As long as mining had been confined within California's own borders, capital had shunned it as a form of investment. Just before the discovery of silver on the Comstock, there had indeed been a movement of urban capital into California water companies, but this had not been accompanied by an equivalent tendency to invest directly in mining, save for a few instances of "capitalists," businessmen, and politicians, becoming interested in quartz mines.

Capital, and especially urban capital, had been very wary of acceding to the appeals of the "honest miner." "Scarcely a merchant in San Francisco could be induced to invest a dollar in any mining enterprise. To do so was to lose *caste*, and endanger one's credit 'on change.'

[20] See *Mining and Scientific Press*, August 17, 1863, August 13, 1864; *Grass Valley National*, October 13, November 7, 1863, February 20, March 8, May 21, 1864; Grass Valley *Daily National*, August 23, October 11, 1864.

Even our foundrymen, could not be induced to invest a dollar, in a business to which they were then, as now, indebted for almost their entire business." [21]

When mining thrust its frontier across the Sierras to a difficult and undeveloped country, and when it changed from the comparatively well-known field of gold to the untried business of exploiting silver veins, then, paradoxically, city capital suddenly became willing to invest. While the snow still lay deep on the mountains, in the winter of 1859-60, a fever for speculation in Nevada mining stocks began. It raged alike in San Francisco and the other commercial cities and in the California mining towns. A telegraph line had recently been strung across the Sierras, and during the winter months it was kept working overtime by eager speculators.[22]

In California vein mining, most of the operations, save for those ill-fated ones at the start of the fifties, had been conducted on a basis of direct personal ownership and responsibility.[23] In Washoe, as Nevada was usually called, the incorporated company very soon became the standard practice, and public sale of shares of stock became universal. By the latter part of 1860, Washoe mining stocks were bought and sold daily at the principal

[21] *Mining and Scientific Press*, February 16, 1863.
[22] Lord, *Comstock Mining*, pp. 77-79; North San Juan *Hydraulic Press*, March 24, 31, 1860.
[23] *Grass Valley Telegraph*, February 12, 1856; Grass Valley *Nevada National*, May 26, 1860.

business houses of San Francisco. Two years later the first stock exchange in the state was established at that city for the primary purpose of facilitating transactions in mining securities, while in 1863 Sacramento and Stockton each acquired a similar institution.[24]

Both the formation of incorporated companies and the public sale of shares of stock were rendered necessary by the high cost and risk involved in developing Nevada mines. In that day of low business morals this type of financial organization frequently lent itself to dishonest manipulation and "wild-cat" speculation, but these dangers were little heeded until several years had passed.[25]

In the meantime every San Franciscan—if not every Californian—who controlled or could borrow any surplus funds greedily invested in Washoe. Men whom one would never suspect of an interest in mining speculations now cast aside their scruples and joined in the pursuit of wealth. For example, the Reverend Henry Durant, the New England clergyman who founded the college that has become the University of California, was the backer of a small quartz mill in Nevada. Easterners and Europeans also, disturbed at the uncertainties of wartime conditions east of the Mississippi, and attracted by the

[24] *Mining and Scientific Press*, November 30, 1860, November 29, 1862, March 23, November 2, 1863; Joseph L. King, *History of the San Francisco Stock and Exchange Board* (San Francisco, 1910), pp. 3-14.
[25] *Mining and Scientific Press*, October 7, 1862, February 23, 1863, July 2, 1864.

high rate of interest in the Far West, began sending their capital to the Pacific Coast for investment.[26]

This rapid development of high finance had its corollary in a revolution in California banking. Prior to the sixties there had not been in California any incorporated general commercial banks. The California state constitution of 1849 had been drafted by men who were deeply impregnated with the anti-bank feeling that was so prevalent in the United States after the panic of 1837. Both the constitution and the early legislation which was based upon it made it almost impossible for a corporation to engage effectively in banking. The obvious need for financial agencies was supplied partly by the express companies, which handled the shipment of gold dust and bullion, and partly by private banks that were owned by individuals in their personal capacity.[27]

Occasionally the private bankers were able to grant assistance to important projects, as in the case of the help given a Columbia water company by the Bank of D. O. Mills & Co. In most instances, however, they were incapable of performing upon a large scale the basic function of a commercial bank: lending credit to business concerns. Neither the size of their capital nor the legal

[26] *Mining and Scientific Press,* December 21, 1861, January 19, February 16, 1863, October 15, November 26, December 17, 1864; *Grass Valley National,* November 17, December 2, 1863; San Francisco *Daily Evening Bulletin,* October 12, November 22, 1864.

[27] "Banks and Banking in California in the Fifties; Early Legislative Prohibitions," *Mercantile Trust Review of the Pacific,* XIII (1924), 118-128.

position of their executives was sufficient to justify major ventures.[28]

With the simultaneous appearance of Washoe silver and local and foreign funds, this primitive financial organization became inadequate. Certain legislative modifications were secured, and in the middle sixties incorporated commercial banks came into being in San Francisco.[29] The largest and most important was the Bank of California, which was incorporated in 1864 with the unprecedentedly large capital of two million dollars. Its president was D. O. Mills, the leading banker of California, but its guiding genius was William C. Ralston, one of the most remarkable financiers America has ever produced.[30]

The intimate relationship between the new type of banking and the mineral industry was soon demonstrated. In 1864 the Bank of California established a branch at Virginia City, on the Comstock Lode. When the Nevada mines and mills needed accommodations, the Bank of California granted them liberally—too liberally. During 1866 and 1867 hard times fell upon the Comstock, and by the summer of the latter year the

[28] Benjamin C. Wright, *Banking in California, 1849-1910* (San Francisco, 1910), pp. 27-28; Ira B. Cross, *Financing an Empire: History of Banking in California* (Chicago, 1927), I, 259-260.

[29] Wright, *Banking*, pp. 47-48; Cross, *Financing*, I, 254-260.

[30] Cf. Cecil G. Tilton, *William Chapman Ralston, Courageous Builder* (Boston, 1935); George D. Lyman, *Ralston's Ring: California Plunders the Comstock Lode* (New York, 1937).

bank, through a succession of foreclosures, had become the owner or controller of most of the important mining properties on the lode. Instead of beating a retreat from the threatening disaster, the bank embarked on a series of major improvements that eventually brought the return of flush days. At one time this great financial house had three million dollars invested in mines, mills, and dependent properties on the Comstock.[31]

Almost simultaneously with the advent of incorporated banks, British banking capital made its entrance into the California field. Several important English banks and agencies were established at San Francisco during 1864 and 1865, and while their contributions to the local money market were not at first as helpful as had been anticipated, they became eventually of great significance as channels through which foreign funds were directed into Pacific Coast investments.[32]

Along with California men and California money, California mining equipment also went eastward across the Sierras. In the early days of the Gold Rush, California learned that she could manufacture better placer implements than she could import, and not long afterwards she made the same discovery in regard to quartz. Iron foundries began producing domestic-made tools at San Francisco in 1849, and during the next eight years foundries and machine shops were established at Sacra-

[31] Lord, *Comstock Mining*, pp. 244-301.
[32] *Mining and Scientific Press*, October 8, 22, 29, 1864, July 8, 1865; San Francisco *Daily Evening Bulletin*, October 12, 1864.

mento, Stockton, Marysville, and some of the chief mining towns.[33]

The long struggle to improve quartz machinery proved an especial boon to the iron workers. It created a large demand for heavy castings and caused the best mechanical minds in the state to devote their time to developing more efficient stamp mills, arrastres, steam engines, and other basic units required for all types of lode mining.[34] When the working of silver veins suddenly became important in 1860, the mechanics who had been trained by the needs of the gold-quartz industry were able to turn to the new field with the confidence that arises from a full knowledge of one's trade.

By midsummer of 1861 nearly one thousand hands were employed in the manufacture of mining equipment at San Francisco alone, and every river and coastwise steamer was departing with a heavy load for Nevada, Mexico, and other silver regions.[35] The rush of business that began in 1860 continued through 1866, in which year the value of the castings produced at San Francisco was only a little less than two million dollars. The greater part of this was destined for the mines. There were at that time thirteen iron foundries and thirty

[33] *Scientific Press*, September 8, 1860; *Mining and Scientific Press*, July 27, 1861; *Sacramento Daily Union*, January 1, 1856; *Sacramento Weekly Union*, January 17, 1857; William P. Blake, *Notices of Mining Machinery and Various Mechanical Appliances in Use Chiefly in the Pacific States and Territories* (New Haven, Connecticut, 1871), p. 2.

[34] *Grass Valley Telegraph*, December 4, 1855.

[35] *Mining and Scientific Press*, July 27, August 17, 1861, April 20, 1863.

machine shops in San Francisco, and twenty-three iron foundries in other parts of the state.[36]

These skilled mechanics and their large manufacturing facilities formed one more link in the chain of factors that were necessary for the rapid expansion of mining during the sixties. Still another link was provided by the presence of men who had had years of experience in the California quartz mines. The exploitation of silver veins, which played so large a role in the new era, presented certain problems that were more difficult than any encountered in dealing with lodes of gold, but with the good practical training that they had received in California, veteran gold-quartz men were able to devise ways of overcoming most of the obstacles.

Perhaps the best illustration of this was the story that lay behind the evolution of the Washoe Pan Process. Almarin B. Paul, a veteran of nine years' experience in mining and milling California gold-quartz, put the first stamp mill into operation on the Comstock in August, 1860. Tests soon convinced him that the problem of dealing with sulphurets and sulphates, while important with gold, was of primary significance with silver. In order to resolve the difficulty he took two complementary ideas that had been developed by two San Francisco foundrymen, added to them a method devised by a quartz mill operator of Sonora, California, and assembled

[36] Blake, *Notices of Mining Machinery*, p. 1; Browne, *Report* (1868), pp. 226-227.

out of them, in the light of his own accumulated knowledge, the Washoe Pan Process. The process proved so successful that it was copied by silver companies the world over.[37]

The expansion of the mining area during the sixties was by no means confined to the Comstock Lode or to silver districts. At the same time that California-trained miners were laying the foundations for the silver industry at Virginia City, similar miners were leading the way into a vast and varied domain that extended from British Columbia into Mexico and from the Rockies to the Pacific.

The rush to Fraser River in 1858 had been the opening move in a great outward thrust of the mining frontier. After the collapse of that initial boom, matters had languished for two years on the Canadian side of the line. Then the pace had quickened when gold was discovered at Cariboo, British Columbia, in 1861, at Kootenai, in the same province, in 1863, and on the Upper Columbia in 1865.[38]

[37] See Paul's controversial correspondence in *Mining and Scientific Press*, March 13, 20, 27, April 3, 1869; also Lord, *Comstock Mining*, pp. 80-83; George D. Lyman, *The Saga of the Comstock Lode, Boom Days in Virginia City* (New York, 1934), pp. 135-137. Unbeknown to Californians, the fundamentals of the process had long been in use in Mexico.

[38] William J. Trimble, "The Mining Advance into the Inland Empire, A Comparative Study of the Beginnings of the Mining Industry in Idaho and Montana, Eastern Washington and Oregon, and the Southern Interior of British Columbia, and of Institutions and Laws Based upon that Industry," *Bulletin of the University of Wisconsin*, no. 638, History Series, III (Madison, Wisconsin, 1914), pp. 176, 182-183, 192, 195.

South of the international boundary, the year 1861 saw the beginning of a great rush into the Nez Percés and Salmon River sections of northern Idaho, and of a lesser movement into eastern Oregon, while the two seasons that followed brought the opening of the rich Boise and Owyhee districts in southern Idaho and of four important mineral regions in Montana.[39]

Further south, and along the line of the Rockies, successful prospecting had been done in Colorado during 1857 and 1858, with the result that during 1859 and 1860 thousands of miners and would-be miners set out with the grim resolve to reach "Pike's Peak or bust." [40] In Utah, Colorado's western neighbor, the Mormons found placer gold as early as 1861. Much prospecting and a limited amount of mining for both gold and silver were done thereafter within that territory.[41]

South of Utah, the working of Arizona's mineral wealth was impeded by the fierce raids of the Apaches, but enough was done in the late fifties and the sixties to extract several million dollars' worth of gold, silver, and copper.[42] New Mexico was similarly retarded by Indian

[39] *Ibid.*, pp. 198-218.

[40] Ovando J. Hollister, *The Mines of Colorado* (Springfield, Massachusetts, 1867), pp. 7-70, 107, 115.

[41] Robert G. Raymer, "Early Mining in Utah," *Pacific Historical Review*, VIII (1939), 81-88.

[42] Eldred D. Wilson, J. B. Cunningham, and G. M. Butler, "Arizona Lode Gold Mines and Gold Mining," *University of Arizona Bulletin*, V, no. 6 (August 15, 1934), pp. 16-17; Frank C. Lockwood, *Pioneer Days in Arizona: From the Spanish Occupation to Statehood* (New York, 1932), pp. 193-200, 214; Robert G. Raymer, "Early Copper Mining in Arizona," *Pacific Historical Review*, IV (1935), 123-130.

forays, and it suffered also from the effects of an invasion by Confederate troops in 1861-62. These hindrances were sufficient to keep gold mining on a limited basis until the middle and later sixties.[43] Meanwhile, impatient Americans turned still further southward to try the potentialities of the mines of Lower California, Sonora, Durango, and the other provinces of northern Mexico.[44]

While these great extensions were being added to the farthest limits of the known mineral area, indefatigable prospectors were filling in the gaps that surrounded the California and Comstock districts. In Nevada the Humboldt District was discovered in 1860 and Reese River in 1862, while near the close of the decade White Pine became for a brief moment a nationwide sensation.[45]

Within the political limits of California, though physiographically divorced from it, a whole succession of mineral districts was opened in the arid, inaccessible tongue of land that lies to the east and southeast of the Sierras. The most important of these were just south of the point where the Sierras pinch against the Nevada line. Here the districts of Bodie and Esmeralda were

[43] Waldemar Lindgren, Louis C. Graton, and Charles H. Gordon, "The Ore Deposits of New Mexico," U. S. Geological Survey, *Professional Paper*, no. 68 (Washington, 1910), pp. 17-18.

[44] *Mining and Scientific Press*, May 11, July 27, 1863, January 21, 1865.

[45] Effie M. Mack, *Nevada, a History of the State from the Earliest Times through the Civil War* (Glendale, California, 1936), p. 215; *Mining and Scientific Press*, October 31, December 5, 1868, March 6, 20, 1869.

organized in 1860, at a time when the surveyors were still undecided as to whether the region belonged to California or to Nevada. Eventually the former was assigned to Mono County, California, and the latter to Nevada.[46]

From Bodie and Esmeralda the chain of mines— mostly silver—followed the curve of the Sierras down to its southernmost point. Thence the bounds of the mineral area jumped across a large expanse of barren ground to the isolated districts near San Bernardino, which is inland some fifty or sixty miles from Los Angeles.[47]

Into all of these new mining regions California sent her men, her money, and her machinery. In all of them California-trained miners passed on to less experienced adventurers the lessons they had learned by trial and error in the Sierra foothills. Californians pioneered the Nez Percés and Salmon River districts in Idaho and did the first prospecting in Montana.[48] "The skill of old Californians was pre-eminent, and everywhere from Cariboo to Owyhee the methods and opinions of Californians were given great respect." [49]

Veterans of the California mines played an important

[46] Joseph Wasson, *Bodie and Esmeralda, Being an Account of the Revival of Affairs in Two Singularly Interesting and Important Mining Districts* (San Francisco, 1878), pp. 4-6, 43-44.

[47] Cf. map of "South-Eastern California Silver Mines," *Mining and Scientific Press*, December 21, 1861.

[48] Trimble, "Mining Advance," pp. 201-203, 206-207, 215.

[49] *Ibid.*, p. 229.

part in uncovering Colorado's mineral wealth at the close of the fifties. Other California veterans are said to have discovered the gold deposits at Pinos Altos, New Mexico, in 1859 or 1860. Still other Californians provided the leadership for organizing the first mining district in Arizona, in 1861. In 1862 Californians were rushing off to British Columbia. From 1862 through 1865 great crowds of them were stampeding now to Boise and the Snake River, now to the San Francisco Mountains of Arizona, now to Colorado, now to Reese River and the Humboldt, and now to Owen's River, Bodie, and Coso along the eastern edge of California itself.[50]

As if not content with making western Canada and the United States their province, the Californians overflowed into Mexico. A steamship line was inaugurated between San Francisco and Mexican west coast ports, and in 1863 every south-bound steamer was sailing with crowded passenger lists and heavy consignments of equipment. In a single district of Lower California, mining men from American California were developing nearly sixty mines. One San Francisco-owned company

[50] Frank Hall, *History of the State of Colorado, Embracing Accounts of the Pre-Historic Races . . . The Original Discoveries of Gold* (Chicago, 1889-90), I, 180-181, 188-192, 209; Lindgren, Graton, and Gordon, "Ore Deposits of New Mexico," p. 18; *Mining and Scientific Press*, June 9, July 23, September 11, 1862, February 16, September 14, 21, 1863, March 5, April 16, May 14, June 25, July 30, 1864, May 27, 1865, February 19, 1867; *Grass Valley National*, August 9, 14, 1862, October 15, 24, December 12, 1863, March 29, 31, 1864.

was reported to have spent $90,000 and another $30,000 in Lower California.[51]

In an attempt to summarize this omnipresent influence of the Californians, a historian of western mining has said:

Whatever elements of population prevailed in one or the other place [in the new mining domain], there was one everywhere present, everywhere respected, everywhere vital—the Californian. To Fraser River, Cariboo, Kootenay; John Day, Boise, Alder Gulch, Helena, went the adopted sons of California—youngest begetter of colonies,—carrying with them the methods, the customs, and the ideas of the mother region, and retaining for it not a little love and veneration. "Idaho," said the [Idaho] *World* [in 1865], "is but the colony of California. What England is to the world, what the New England states have been to the West, California has been and still is to the country west of the Great Plains. Her people have swept in successive waves over every adjacent district from Durango to the Yellowstone. She is the mother of these Pacific States and Territories." [52]

How was any young western state, itself not two decades old, able to do so much? The answer must be that California was unlike any previous commonwealth that had existed west of the Alleghanies. In all earlier western states and territories an inadequate supply of labor, capital, and home-produced manufactured goods had been fundamental limiting conditions.

---

[51] *Mining and Scientific Press*, May 11, 18, July 27, December 7, 1863, October 1, 1864.

[52] Trimble, "Mining Advance," p. 141.

California's ability to do so much was primarily the result of her having come into being as a mining rather than as an agrarian community. Mineral wealth attracts into a new country a great many people within a very short period of time. They come thither for the specific purpose of making a quick fortune, and as soon as the opportunity for doing so seems to be fading, they are ready to move on to a still newer mineral region. Mineral wealth tends also to build up accumulations of capital, and to bring into being heavy industry, transportation facilities, and impressive cities that would have arisen only after years of slow growth in an agricultural country.

This was the condition of California at the close of the fifties. She had the manpower, the capital, the manufacturing resources, and the experienced mining leaders. And she had a huge fund of disappointed hopes and restless ambition.

Had she been an eastern state, her energies and abilities would have been absorbed into the exhausting struggle of the Civil War. That great conflict, however, did not deeply affect the West that lay beyond the Great Plains. On the contrary, emigration to California and Nevada from some parts of the "States" may actually have been stimulated by the troubled conditions of wartime. The west was left undisturbed in its task of extending the nation's settled limits, while the East fought the nation's battles. As the *Sacramento Union* remarked:

It will be seen that while we are defeating an iniquitous scheme to tear away from the Union the States of the South, we have the energy and the population to secure and develop [in the West] a region greater in extent than the theater of rebellion. . . . Such is the evidence of [America's] imperial, exhaustless power.[53]

Thus California was left free to employ her men and treasure in pioneering an empire that extended from southern Canada into northern Mexico, and from the Great Plains to the Pacific. When the returns from the census of 1870 led someone to ask why California had not grown more rapidly during the decade, the editor of the *Alta California* replied:

We have had not only to found a State here but have brought into life other States and Territories. Nevada is the child of California. Oregon has received considerable accessions to its population from us. Idaho and Arizona have been mainly settled by Californians. The work in which we are in fact engaged is the building up and peopling of half of a great continent.[54]

[53] *Sacramento Weekly Union*, December 17, 1864.
[54] San Francisco *Daily Alta California*, February 3, 1872.

XII

THE REGULATION OF SOCIETY
1848-1873

When the California veterans started out from the Golden State into the new frontier, they not only carried with them the technical knowledge of how to exploit minerals and build mining camps, but they also bore in their minds an established attitude towards the government of society in a new mining country. For ten or a dozen years California had been experimenting with the handling of social problems, and by the time of the great exodus she was ready to pass on to newer commonwealths the lessons she had learned—for better and for worse.

There had been two main tasks: first, the organization of society so as to prevent and control crime; second, the regulation of ownership of mining claims. In dealing with both the Californians had to bring out of the chaos of the Argonauts' motley origins a semblance of order and impromptu discipline.

Historians have written much about the American pioneers' instinct for spontaneous organization and self-government. Institutionally the trait has been traced back to the practices of the dissenting religious sects in England. Its development in America has been followed along paths that are marked by such milestones as the Mayflower, Watauga, and Cumberland Compacts, the Regulators' and land-claim associations, and the movements for statehood in the territories of the first American "West." [1]

On a small scale this ability for self-organization showed itself in California in the hundreds of mining companies. The first of these were formed in 1849 before the Argonauts left their homes in the East. It was customary for groups of intending gold seekers to unite their efforts in joint-stock associations, for mutual protection in the wild country to which they were going and to facilitate a sharing of expenses. Eastern investors often furnished capital to such companies in return for shares of stock.[2]

Almost all of these companies disintegrated soon after reaching California. The elaborate charters and constitutions under which most of them were supposed to operate had been drawn up by persons ignorant of California conditions. They generally proved unworkable when put into force.[3]

[1] A convenient summary may be found in Mary F. Williams, *History of the San Francisco Committee of Vigilance of 1851, a Study of Social Control on the California Frontier in the Days of the Gold Rush*, University of California Publications in History, XII (Berkeley, California, 1921), pp. 8-18.

[2] Mulford, *Prentice Mulford's Story*, p. 8. An example was: Hartford Union Mining and Trading Company, *Articles of Association and By-Laws of the Hartford Union Mining & Trading Co. Adopted January 19th, 1849* (Hartford, Connecticut, 1849).

[3] San Francisco *Daily Alta California*, July 27, 1851; Woods, *Sixteen Months*, pp. 170-171, 181.

They were soon succeeded by the companies which the miners formed to carry out projects too large for a single man or a partnership. River mining and water systems were especially apt to lead to organizations of this type. When a miner joined one, he not only agreed to invest his labor in the common effort, but also to obey the orders of the elected officers, to be satisfied with only his proportionate share in the profits, and not to absent himself from work without authorization, save in cases of serious illness.[4]

In commenting on these companies, an observant Scotchman remarked that the Americans showed a greater organizational ability than the other nationalities in the mines. He said:

In this respect the Americans had a very great advantage, for, though strongly imbued with the spirit of individual independence, they are certainly of all people in the world the most prompt to organise and combine to carry out a common object. They are trained to it from their youth in their innumerable, and to a foreigner unintelligible, caucus-meetings, committees, conventions, and so forth, by means of which they bring about the election of every officer in the State, from the President down to the policeman; while the fact of every man belonging to a fire company, a militia company, or something of that sort, while it increases their idea of individual importance, and impresses upon them the force of combined action, accustoms them also to the duty of choosing their own leaders, and to the necessity of afterwards recognizing them as such by implicit obedience.[5]

[4] See above, p. 60.
[5] Borthwick, *Three Years*, p. 369.

It was fortunate that the Americans had this inbred ability for self-organization. At the time of the gold discovery the province was limping along under a weak regime that was headed by a few army officers at Monterey, but which operated, for civil purposes, through a modified version of the governmental agencies inherited from Mexico.[6] The people were restive under this unsatisfactory rule and held several meetings in 1848 and 1849 to debate the need for a genuine territorial government.

Congress, however, was deadlocked over the slavery issue and hence was unable to provide for California. Impatiently, and under the unexpected leadership of their soldier governor, the American and native Californian populations finally took upon their own heads the responsibility of drawing up a state constitution and electing a state government. Without waiting for the approval of Congress, they put the new administration into power in December 1849.[7]

In view of the social confusion of the Gold Rush period and the diversity of backgrounds of the citizens of the new commonwealth, this was in itself a notable achievement. Yet it was not a sufficient answer to the demands of the time. As one might expect, there was a lapse of several months before the machinery of con-

[6] This regime is described in documents printed in: "California and New Mexico," *House Exec. Doc.*, 31 Cong., 1 sess., no. 17, pp. 597-598, 765-766, 772-777.
[7] Bancroft, *Works*, XXIII, 284-336.

stitutional government was in full operation. Not until April or May of 1850 were the larger towns incorporated and the counties and townships supplied with officials.[8]

There was an even longer interval before a working system of law was provided. It had been hard enough to discover the contents and meaning of the Mexican codes under which the province was, in theory, governed prior to December 1849, but because of a prolonged delay in publication, it was virtually impossible to find out the provisions of the new state statutes. In October 1851 the leading newspaper declared: "For the last two years the anomaly has been presented in this State of a people bound by a set of laws, of whose purport they have almost been in an utter state of ignorance." [9]

The slow start of the new agencies resulted in a continuance for many months of the condition of minimum government which had prevailed under the preceding military-civil regime. The absence of effective rule, in turn, placed directly upon the people, in their primary capacity as the "body politic," the responsibility for preserving a semblance of order.

The men of the Gold Rush population responded well to the challenge. They did not have to deal with crime of serious proportions in 1848 or the first half of 1849, but thereafter they had to face the defiance of as dan-

[8] Williams, *History of the San Francisco Committee*, p. 117.
[9] San Francisco *Daily Alta California*, October 22, 1851.

gerous a group of malefactors as has ever threatened any society.

California was so unfortunate as to receive a large number of ready-made convicts from Britain's Australian penal colony. She was unfortunate also in acquiring from all parts of the world a miscellaneous assortment of riffraff, to which home-grown recruits were constantly being added during the early and middle fifties as the declining placers and the gambling houses sent forth their disappointed and sometimes desperate thousands.[10]

Furthermore, California did not prove a friendly home to all of the foreign immigrants who came to it. The Latin Americans, French, and Chinese were all subjected to organized persecution. The experience of participating in persecution doubtless did much in itself to accustom shallow or cruel men to the commission of lawless acts, while the brutalities perpetrated on the Sonorans gave many members of that half-civilized race what little excuse they needed for turning to a life of crime.[11]

The threat to society from evil doers made itself felt at two points: in the mineral region and in the commercial cities. In the former, legally constituted governmental agencies were at a minimum. New mining communities were constantly springing into existence in

[10] Williams, *History of the San Francisco Committee*, pp. 86, 120-121.
[11] Guinn, "Sonoran Migration," Historical Society of Southern California, *Annual Publications*, VIII, 33-36.

remote canyons where none but the Indian had dwelt before. It was too much to expect that the feeble provincial or state regime would quickly provide each one with law-enforcement machinery.

With the Gold Rush mob left thus to its own resources,

a state of things little short of anarchy might have been reasonably awaited.

Instead of this, a disposition to maintain order and secure the rights of all, was shown throughout the mining districts. In the absence of all law or available protection, the people met and adopted rules for their mutual security—rules adapted to their situation where they had neither guards nor prisons, and where the slightest license given to crime or trespass of any kind must inevitably have led to terrible disorders. Small thefts were punished by banishment from the placers, while for those of large amount or for more serious crimes, there was the single alternative of hanging.[12]

The usual procedure was for the whole camp to assemble as soon as word had been passed from mouth to mouth that a crime had been committed and the supposed criminal apprehended. Desiring to preserve as nearly as possible the legal forms to which they had been accustomed, the crowd would usually see to it that legal counsel was appointed to present both sides of the case. The trial would take place before the people collectively —the "body politic." Sometimes the people would serve directly as a mass jury. More commonly they would

[12] Taylor, *Eldorado*, p. 100.

elect a judge and jury from amongst themselves. If the accused were found guilty, the sentence of whipping, banishment, or hanging would be carried out immediately, since there was no jail in which to confine the convicted man. Their job done, the crowd would "set up the drinks all around" and then return to work.[13]

Many camps, especially after the opening of the mining season of 1849, set up a more permanent law-enforcement mechanism by electing a magistrate and a sheriff. In deference to Mexican terminology, the former was called an *alcalde* prior to the establishment of state government. Subsequently he was given the normal Anglo-Saxon title of justice of the peace. His function was to handle civil and minor criminal cases and to preside over major criminal trials if the camp so desired.[14]

Clearly these simple forms of "popular justice" were little more than lynch law dressed in the full trappings of judge-and-jury trial,[15] yet it is hard to see what other recourse there was during the first few years after the gold discovery. As the *Alta California*, the state's best known newspaper, explained: "Lynch law is not the best law that might be, but it is better than none, and so far as

[13] In general, see Williams, *History of the San Francisco Committee*, pp. 76-81, and Charles H. Shinn, *Mining Camps, A Study in American Frontier Government* (New York, 1885), pp. 177-180.

[14] Williams, *History of the San Francisco Committee*, pp. 81-82; Shinn, *Mining Camps*, pp. 182-205; Taylor, *Eldorado*, pp. 101, 263; Marryat, *Mountains and Molehills*, pp. 326-328.

[15] Lynch law is "the act or practice by private persons of inflicting punishment for crimes or offenses, without due process of law," *Webster's New International Dictionary*.

benefit is derived from law, we have no other here." [16]

The danger in lynch law was, of course, that the crowd might be swept off its feet by a surge of prejudice or momentary anger, and thus hang an innocent man, or that the crowd might be swayed by a wave of maudlin sentiment and release a criminal. Reviewing the situation in midsummer of 1852, the *Alta California* pointed out that the slow materialization of the state and local governments had gradually removed the excuse for lynch law. By that time, the *Alta California* felt, the law-enforcement machinery was too complete for popular action to be justifiable any longer: "Society, too, is becoming organized, and the population so much increased that a criminal has small chance of escaping detection and punishment compared with what he had two or three years since." [17]

The difficulty was that once started, the practice of impromptu justice was hard to stop. Two and a half years later another influential journal, the *Sacramento Union*, complained:

There is too ready a disposition manifested in the State to resort to lynch law—to inflict summary justice upon offenders. Four years ago there may have existed an absolute necessity for courts of, and executions by the people, in their primary capacity. It cannot justly be said that any such necessity now exists.[18]

[16] San Francisco *Weekly Alta California*, March 8, 1851.
[17] San Francisco *Daily Alta California*, August 12, 1852.
[18] *Sacramento Daily Union*, January 11, 1855.

According to a San Francisco newspaper, during the year 1855 forty-seven men were executed illegally in California, as against only nine lawful executions. As one would expect, most of the lynchings took place in the mines and other rural areas, because in those regions lawlessness continued to be a frequent problem for several years after a measure of peace had come to the cities. The homicide statistics for 1855 revealed only fifteen violent deaths in San Francisco County, seven in Sacramento County, and eight and four, respectively, in the counties in which Marysville and Stockton were located. Of the counties devoted to mining, on the other hand, Calaveras had thirty-two homicides, El Dorado twenty-six, Amador twenty-one, Siskiyou twenty, and Tuolumne nineteen.[19]

Even in the cities there were periodic spasms of backsliding into what were considered intolerable conditions. In the cities, and especially in San Francisco, the situation was complicated by the fact that although complete governmental machinery existed, it was so ridden by corruption and weakness as to be inadequate for the task of maintaining order.[20] It was this circumstance which gave rise to the several great vigilance committees.

A "lynching party" was usually a mob that came hastily together to deal with a particular crime. A vigilance committee was a much more formal organization.

[19] Ibid., December 31, 1855, clipping San Francisco Chronicle.
[20] Hittell, History of San Francisco, pp. 172, 241-242.

Realizing that they must be prepared temporarily to supersede the legitimate municipal government and to rule the varied population of their community, the leaders of the great vigilance committees were careful to shape the structure and select the personnel for their association before beginning their dread labor. Members were assigned to definite tasks in military fashion. Solemn deliberation preceded important actions, and punishment was meted out only after a trial that was made closely to resemble legitimate processes.[21]

The greatest of all vigilance committees were those that took over the government of San Francisco for several months in 1851 and 1856.[22] In both cases the committee roster included the leading private citizens, and in both the net effect of the vigilance work was to purge the city of its most dangerous elements. A few notorious criminals were hanged, a larger number were sent into exile, and a still larger number were frightened into flight by the mere presence of summary justice. The second committee achieved a more permanent reform by setting up a non-partisan citizens' association which for many years dominated the local elections.

The example of San Francisco was widely copied.

[21] *Ibid.*, p. 261.
[22] See Williams, *History of the San Francisco Committee*, pp. 163-390; Bancroft, *Works*, XXXVII, *passim;* Stanton A. Coblentz, *Villains and Vigilantes: The Story of James King of William and Pioneer Justice in California* (New York, 1936); James A. B. Scherer, "*The Lion of the Vigilantes,*" *William T. Coleman and the Life of Old San Francisco* (Indianapolis and New York, 1939).

All of the commercial cities, several of the larger towns, and some of the camps had vigilance committees at one time or another during the fifties. In a few unstable areas it was even necessary to resort to vigilantes in the sixties and seventies. Some of the "country committees," as Bancroft termed them, were worthy of a place beside the much praised San Francisco organizations, but others were little more than lynching mobs masquerading under the more respectable title.[23]

Beyond California's boundaries there were a few far western vigilance committees and lynching parties during the fifties. There were many more during the sixties and early seventies, when the extension of the mining frontier carried so many Californians into Nevada, Idaho, Montana, Oregon, and Arizona.[24] The appearance in these areas of both vigilance committees and lynching parties that claimed to be vigilance committees was the result of the tendency of the new commonwealths to reënact the unsettled conditions that had prevailed in California a decade earlier.

Washoe, for example, was said to be at the start of the sixties what California had been at the start of the fifties: a gathering place for evil doers who would not be tolerated elsewhere. Idaho, similarly, repeated California's experience of struggling against an absence of courts of law and a chaotic government. In short, California's

[23] Bancroft, *Works*, XXXVI, 441-514; Williams, *History of the San Francisco Committee*, pp. 374-387, 406-407.
[24] Bancroft, *Works*, XXXVI, 593-749.

children suffered from the same ills which had once so disturbed the parent state, and in their travail they turned to the same impromptu remedies that had in California so often taken the place of the legitimate processes of a normal society.[25]

[25] *Ibid.*, XXXVI, 601-602, 655. Colorado, on the other hand, seems to have done better. Lynn I. Perrigo, "Law and Order in Early Colorado Mining Camps," *Mississippi Valley Historical Review*, XXVIII (1941-42), 41-62.

XIII

THE LAW OF THE MINES
1848-1873

More important than the attempts to punish crime was the constructive work of the miners in evolving a set of rules by which the ownership of mining claims could be regulated. This was one of the earliest legal problems encountered in California, and it was one for the solution of which previous American experience could offer little guidance.

There was no significant body of American mining law in 1848. The gold regions of Georgia, North Carolina, and the other southern states had produced nothing more than a few simple, local rules. The lead mines of Iowa had done better. In the Dubuque district the miners had assembled and selected a committee to draw up regulations which would specify the number of square yards of mineral ground each man might occupy, and the procedure for settling disputes.[1]

The Iowa method indicated the course that was to be followed in California. Those same miners' meetings that provided the mechanism for dealing with criminals during 1848, 1849, and 1850 served also as the means for establishing a code of mining law.

[1] Shinn, *Mining Camps,* pp. 39, 44-45.

When mining began in California, almost all of the gold deposits were on public land that had recently been acquired by the United States from Mexico.[2] Ever since the days of the land ordinance of 1785, it had been the policy of the United States, as expressed in many congressional statutes, to reserve the mineral-bearing parts of the public domain from general sale or preëmption.[3] From 1807 to 1846, in dealing with the mineral lands of the Middle West, Congress had tried to maintain a system of leasing, but by the latter date the attempt had become untenable because of the refusal of the miners to continue paying rent to the federal officers. During the two years that preceded Coloma, Congress acknowledged the failure of the leasing system by passing four acts aimed at transferring the Midwestern reserved lands to private ownership. Congress did not, however, alter its reservation policy insofar as it applied to other parts of the country.[4]

Marshall's discovery, therefore, occurred at a time when there were no well-developed American mining codes and no federal regulations other than the general policy that mineral lands not otherwise provided for were not subject to sale or preëmption.[5] Left thus upon their

[2] The chief exception, the Mariposa Estate, is discussed in Chapter 17.
[3] Gregory Yale, *Legal Titles to Mining Claims and Water Rights in California, under the Mining Law of Congress, of July, 1866* (San Francisco, 1867), pp. 325-330. This book was the first great treatise on mining law in the West.
[4] *Ibid.*, pp. 337-338; Shinn, *Mining Camps*, pp. 40-41.
[5] Shortly after the gold discovery the American military governor of California abolished the Mexican law concerning rights to mineral land.

own resources, the miners of the early years turned for advice to those same Latin American veterans who were so often their instructors in the art of mining.

The Latin Americans had been trained under the mining ordinances of Spanish America. Those ordinances, in turn, were the New World adaptation of the codes which had been developed in Europe during several centuries of experience with handling mining disputes and regulating mining practice. A Californian said of them: "These [European and Latin American] regulations are founded in nature, and are based upon equitable principles, comprehensive and simple, have a common origin, [and] are matured by practice." [6]

In other words, the codes were the product of the common experience of miners in several parts of the world in dealing with the typical problems that were likely to arise wherever men sought mineral wealth. Most of them could be used as satisfactorily in California as they had been in Europe and New Spain. By applying fragments of them in piecemeal fashion, in order to settle specific disputes as they arose, the early miners slowly became familiar with the whole body of this long-established law. Because it suited their needs, they gradually took over into their daily practice the greater part of the system, after making such modifications as were

Cf. proclamation dated Monterey, February 12, 1848, "California and New Mexico," *House Exec. Doc.*, 31 Cong., 1 sess., no. 17, pp. 476-477.
[6] Yale, *Legal Titles*, p. 58.

deemed necessary by the local peculiarities of the California setting.[7] In California the codes seemed to work so well that it became customary to ascribe to them an exclusively Californian origin.[8]

The process of adoption took many months, perhaps several years, and drew not only upon Spanish and Spanish American experience, but also upon the experience of English miners who began arriving in 1849 from the ancient tin and lead districts of Cornwall, Devon, and Derby. As fully developed and used in California,

most of the rules and customs constituting the code, are easily recognized by those familiar with the Mexican ordinances, the Continental Mining Codes, especially the Spanish, and with the regulations of the Stannary Convocations among the Tin Bounders of Devon and Cornwall, in England, and the High Peak Regulations for the lead miners in the county of Derby.[9]

The medium through which 'the codes passed into California usage was the miners' meeting. In 1848 this seems to have been an informal and irregular affair. When a dispute arose between two men who claimed the same piece of ground, the friends of one of the disputants would circulate a request for the miners of that

---

[7] Cf. Henry W. Halleck, "Introductory Remarks, by the Translator," in J. H. N. DeFooz, *Fundamental Principles of the Law of Mines* (San Francisco, 1860), p. vii.

[8] A tendency well illustrated by Senator Stewart's famous speech in 1866. *Congressional Globe*, 39 Cong., 1 sess., pp. 3225-3229.

[9] Yale, *Legal Titles*, p. 58.

particular camp to assemble and settle the argument by discussion and majority vote.[10]

During 1849 and the subsequent years, a more permanent arrangement was adopted.[11] It became customary to hold at each camp or in each local territorial unit—such as a gulch or a portion of a river valley—an organizational meeting open to all the miners working there. At the meeting the camp or local unit would be formally declared a mining "district," and a committee would be elected to draw up, for submission to the meeting, a set of regulations governing all claims in the district.

Specific details of the regulations were likely to vary widely between districts, but the essentials were the same, since all had their eventual place of origin in Spanish American and European practice. In all of the codes the fundamental principle was that the man who discovered a tract of auriferous ground was entitled to exploit it. This privilege lasted so long as the discoverer continued actively to work upon his tract. In other words, "The right of property in mines is made to depend upon *discovery* and *development;* that is, *discovery* is

[10] Shinn, *Mining Camps*, pp. 122-128.

[11] The following description of the adoption and contents of the codes is drawn from Yale, *Legal Titles*, pp. 73-84. It represents the matured practice. For a description of a miners' meeting at work, see Borthwick, *Three Years*, pp. 153-157. For typical codes, see Shinn, *Mining Camps*, pp. 237-258, and the same author's *Land Laws of Mining Districts*, Johns Hopkins University Studies in Historical and Political Science, 2nd series, XII (Baltimore, 1884).

made the source of title, and *development*, or *working*, the condition of the continuance of that title." [12]

No one might hold more than one such claim by this process of "locating," but it was usually permissible to add to one's holdings by buying out other locators. The discoverer of a new gold district was customarily allowed to hold twice as much as the later arrivals. There was always a statement as to the maximum number of feet or yards in length and width allowed per claim, and requirements were included concerning the setting up of notices and markers to indicate claim boundaries.

In an attempt to avoid misunderstandings as to ownership, a district recorder was usually elected. It was his duty to make a descriptive list of all claims and register all transfers of title. Realizing, however, that disputes were bound to arise, a simple method of arbitration was set up in most of the later codes. In the earlier ones it seems to have been expected that the miners of the district would assemble and from amongst them a jury would be chosen to settle each case.[13] In the later codes the procedure was more definite. Sometimes it was provided that the parties to each quarrel must select a temporary board of arbitrators. Sometimes the general miners' meeting elected a permanent committee to han-

[12] Halleck, "Introductory Remarks," *Fundamental Principles*, p. vii.
[13] Cf. Borthwick, *Three Years*, pp. 153-155; Marryat, *Mountains and Molehills*, p. 240; *Sacramento Weekly Union*, April 23, 1853.

dle all cases. In either case the arbitral tribunal was always empowered to make a decision that was binding.

Well aware of the mutability of mining conditions, and recognizing that law must be in consonance with the prevailing situation, the miners generally wrote into the codes a provision for subsequent revision. Usually a majority vote of a duly summoned mass meeting was the only requirement for approval of changes.

Collectively these several regulations and the machinery for applying them came within a few years to constitute the customary law of mines in California. They had the same basis as the English common law: universal acceptance and use rather than statutory enactment. They were adopted because they were the readiest solution to the pressing need for "clear and well-defined rules . . . applicable to the new conditions." [14]

Contemporaries were agreed in saying that in general this customary law proved satisfactory. One English observer declared that "it is astonishing how well this system works," [15] and another said of the settlement of a mining dispute, "I must say I never saw a court of justice with so little humbug about it." [16] According to Chief Justice Sanderson of the state supreme court, the miners' rules "were few, plain and simple, and well understood by those with whom they originated. They

[14] *Morton* v. *Solambo Copper Mining Co.,* 26 Cal. 527, 533.
[15] Marryat, *Mountains and Molehills,* p. 239.
[16] Borthwick, *Three Years,* p. 155.

were well adapted to secure the end designed to be accomplished." [17]

Like all bodies of common law, California mining jurisprudence was a growing rather than a static thing. In several respects it underwent great changes during the first ten or fifteen years of its use. One of the most important had to do with remedying the lack of uniformity between the regulations of different districts.

Since each district was a completely autonomous unit, and since the number of districts was very large,[18] it was inevitable that diversity of practice should develop when the traditional Spanish American and European codes were put into operation. By the spring of 1851—after the codes had had at least two and a half years in which to take on local peculiarities—men were becoming convinced of the desirability of greater homogeneity.[19]

In the quartz industry there was at first an attempt to meet the need by formulating a system of rules that would apply throughout the state. A convention was held at Sacramento in July 1851, to draw up "simple rules and regulations [which could serve] as a basis for

---

[17] *Morton v. Solambo Copper Mining Co.,* 26 Cal. 527, 532 (October term, 1864). Note that the codes also regulated the use of water from streams for mining purposes. The basic principle was that the first person to appropriate (i.e., divert) water acquired a possessory right to it— a principle of great importance in irrigation law today. Samuel C. Wiel, *Water Rights in the Western States, The Law of Prior Appropriation of Water* (3rd ed., 2 v.; San Francisco, 1911).

[18] Estimated at about 500 in 1866. Browne, *Report* (1867), p. 226.

[19] San Francisco *Daily Alta California,* July 14, 27, August 5, 1851.

the action of particular precincts or localities." [20] There is no evidence to show that the resulting recommendations were adopted by the local districts. Two years later the legislature was asked to summon an official state convention, but it rejected the proposal, on the grounds that state-wide action was too ambitious for the times. The legislature expressed the opinion that a state convention, be it concerned with vein or with any other type of mining, would do more harm than good unless it were preceded by a prolonged period of careful, officially sponsored law-making by the people of the smaller territorial units, such as the townships and counties. Upon the basis of the collected enactments of these smaller subdivisions, the legislature felt, it might then become possible to attempt legislation by a miners' assembly representing the whole state. [21]

In actuality, this was essentially the practice that was followed in regard to quartz laws, although without the aid or sanction of the legislature. By way of preparation for the Sacramento meeting of 1851, the quartz men of Tuolumne County had convened a county conference of their own and had drafted a tentative code for the consideration of the state body. [22] Almost simultaneously an assembly of quartz miners was meeting in

[20] Ibid., July 15, 1851.
[21] Select Committee . . . on Calling a Miners' State Convention, "Majority and Minority Reports," California Assembly, Document, 4 sess., 1853, no. 35.
[22] San Francisco Daily Alta California, July 11, 1851.

Mariposa County, where they adopted a series of resolutions which were intended to govern all present and future lode locations in the county.[23]

A year and a half later the Nevada County quartz men also resorted to a county meeting, because they, too, felt the need for greater uniformity. In 1859 Sierra County followed suit, while in the meantime the Tuolumne miners had adopted laws that were to prevail throughout their county.[24]

This tendency to substitute the county for the much smaller district was in itself a sign of progress, but the advance did not stop there. The Nevada County quartz laws were gradually recognized as being so satisfactory a model that they were "generally adopted as, or made the basis of, the mining laws of other districts throughout the State." [25]

By this process of local action a greater degree of standardization was gradually achieved in the quartz industry. In placer mining uniformity was a more elusive goal. The number of miners involved was much larger, and there were too many different types of placer diggings, each one of which had its peculiar features. Basic conditions, also, were subject to greater change than was true with quartz veins. In particular, as the shallow placers lost their original richness, the rules governing

[23] *Ibid.*, June 30, 1851.
[24] *Ibid.*, October 28, 1852; North San Juan *Hydraulic Press*, June 18, 1859; Yale, *Legal Titles*, pp. 73-74.
[25] *Mining and Scientific Press*, December 23, 1865.

the exploitation of them had to be revised drastically, while simultaneously the development of deep claims was giving rise to an entirely new set of laws. It is, therefore, understandable why the placer miners failed to make an advance comparable to that of the quartz men.

Despite this handicap, the worst aberrations in the regulations seem to have been gradually eliminated during the first fifteen years of California mining history. In 1864 Chief Justice Sanderson said of the mining codes generally: "These customs and usages have, in progress of time, become more general and uniform, and in their leading features are now the same throughout the mining regions of the State." [26]

Alongside the progress toward greater uniformity should be placed the changing practice in regard to mining litigation. As first established, the codes stood outside the sphere of the legal system provided by the new state constitution. If for any reason a suit involving mining property had been brought in a state court rather than before a local tribunal, the judge would have found himself trying a case that was governed by "laws" that were based neither upon statutes nor upon normal legal precedents. Realizing that eventually mining cases would find their way into the officially constituted courts, Stephen J. Field, then a young state assemblyman from Yuba County, later Chief Justice of the California

[26] *Morton* v. *Solambo Copper Mining Co.,* 26 Cal. 527, 533.

Supreme Court and Associate Justice of the United States Supreme Court, induced the legislature of 1851 to vote that in justices' courts,

in actions respecting "Mining Claims," proof shall be admitted of the customs, usages, or regulations established and in force at the bar, or diggings, embracing such claim; and such customs, usages, or regulations, when not in conflict with the Constitution and Laws of this State, shall govern the decision of the action.[27]

No provision was made respecting trial in the higher courts. When, however, cases began to pass up through the regular legal channels to the state supreme court, that august body declared:

Courts are bound to take notice of the political and social condition of the country which they judicially rule. In this State the larger part of the territory consists of mineral lands, nearly the whole of which are the property of the public. No right or intent of disposition of these lands has been shown either by the United States or the State governments, and with the exception of certain State regulations, very limited in their character, a system has been permitted to grow up by the voluntary action and assent of the population, whose free and unrestrained occupation of the mineral region has been tacitly assented to by the one government [the United States], and heartily encouraged by the expressed legislative policy of the other [California]. If there are, as must be admitted, many things connected with this system, which are crude and undigested, and subject to fluctuation and dispute, there are still

[27] *Statutes of California*, 2 sess., 1851, p. 149 (section 621 of the Civil Practice Act). Cf. Carl B. Swisher, *Stephen J. Field, Craftsman of the Law* (Washington, 1930), p. 55.

some which a universal sense of necessity and propriety have so firmly fixed as that they have come to be looked upon as having the force and effect of *res judicata*.[28]

In other words, in the absence of contrary instructions from Congress and the legislature, the courts were willing to recognize the inescapable fact that the people had adopted laws of their own and were governing all mining operations in accordance with them. That being the case, and with the reasonable supposition that the legislature, at least, approved, the courts were prepared to accept the miners' rules as the law governing miners' cases. As Field later expressed it, after he had gone from the California bench to Washington:

These regulations and customs were appealed to in controversies in the State courts, and received their sanction; and properties to the value of many millions rested upon them. For eighteen years—from 1848 to 1866—the regulations and customs of miners, as enforced and moulded by the courts and sanctioned by the legislation of the State, constituted the law governing property in mines and in water on the public mineral lands.[29]

The miners' jurisprudence thus acquired an official standing in the state courts, and many cases were submitted to the legitimate tribunals for judgment. That does not mean that the local juries and boards of arbitration ceased to exist. On the contrary, the latter offered

[28] *Irwin* v. *Phillips*, 5 Cal. 140, 146 (January term, 1855).
[29] *Jennison* v. *Kirk*, 98 U. S. (8 Otto) 453, 458 (1878).

the advantages of settlement on the spot, by men familiar with the local customs and rules, "in a summary manner, free from complicated forms and intricate proceedings, with little expense, preserving substantial justice, and beyond the reach of corrupt influences";[30] hence "the miners themselves are the first to institute them, and to abide by their judgments," a student of mining law declared in 1867.[31]

Cases were arising, however, in which large sums were at stake and complicated legal questions at issue. It is understandable that such suits should have been transferred to the legitimate courts. How extensive the shift was from the one type of tribunal to the other, it is impossible to say. Some indication may be gained from the fact that mining cases were included in every volume of the California supreme court reports from 1853 to 1870. It is also instructive to note that so vigorous a lawyer as William M. Stewart, the future Nevada senator, could find employment in mining litigation, in the latter half of the fifties, in towns like Nevada City, Nevada County, and Downieville, Sierra County.[32] It seems reasonable to assume that throughout the greater part of California the tendency to resort to the legally constituted tribunals was on the increase as the fifties marched onward into the sixties.

[30] Yale, *Legal Titles*, p. 83.
[31] *Ibid.*
[32] *Representative and Leading Men of the Pacific: Being Original Sketches of the Lives and Characters*, ed. by Oscar Shuck (San Francisco, 1870), pp. 637-638.

When the state courts adopted the miners' codes, they took over a body of law that had many imperfections. Two of the leading defects were well stated by a mining-town newspaper in 1859: "A growing embarrassment to the mining interest is the loose manner in which the laws of mining districts are framed, and the not over nice definition of claim boundaries until a rich strike has been made." [33]

In commenting on this assertion, another mining-town journal remarked that much litigation could be avoided if the several districts in each county or township would agree on a standard and more careful system of surveying and marking the boundaries of claims.[34] Other contemporaries pointed out that the negligence in delimiting claims combined with the inadequacies of the laws themselves to make possible a widespread resort to fraud. A species of legal blackmail was common in the later years: when a long-suffering miner began at last to strike pay dirt, he would find his neighbors suddenly stretching their boundaries onto his ground, or else he would find himself sued by some prospector who had occupied that claim long before but had abandoned it.[35] In such cases the hope of the offending parties was not that they could win in a legal trial, but rather that they

[33] *Sacramento Weekly Union*, May 21, 1859, clipping Downieville *Sierra Democrat*.

[34] North San Juan *Hydraulic Press*, May 21, 1859, editorial remarks concerning the same clipping.

[35] *Grass Valley Telegraph*, May 11, 1854; San Francisco *Daily Alta California*, July 4, 1853, February 23, 1872; *Scientific Press*, March 26, 1870.

could make themselves sufficiently obnoxious to be "bought off."

A further difficulty was that despite the minimum-work stipulations of the codes, it was not uncommon for men to hold claims without performing any significant amount of labor on them. The aim was, of course, purely speculative: to hold until a good selling price could be obtained.[36]

Most of these problems could have been eliminated by tightening up both the laws and the enforcement of them. There seems, however, to have been an increasing amount of legislative inertia after the passing of the flush days. At Dutch Flat, for example, the local editor said that the revision of the laws of his district had failed to keep pace with the change in the character of mining, and that instead of summoning a miners' meeting to correct the deficiencies, it had been the "uniform custom for years, entirely to disregard the former code of mining laws." [37]

One may sum up these criticisms by saying that as law-makers the miners were frequently neither sufficiently careful nor up-to-date, and that as real estate operators they were generally lax. Their failings on both scores were much more productive of costly debate when cases were being tried before distant state courts, than when all disputes were settled on the spot and by

[36] San Francisco *Daily Alta California*, July 27, 1851; Grass Valley *Nevada National*, November 12, 1859.

[37] *Dutch Flat Enquirer*, August 25, 1866.

the judgment of local men who knew the materials with which they were dealing.

Just as this system of law was completing a full decade of development in California, the advance began into the vast domain that extended from the Great Plains to the Pacific. Had they wished to, the pioneers of the new empire could have turned away from California's experience. They could have sought guidance from the English common law and from Congress. Instead, as the Public Lands Commission later discovered:

Since the developers of the great precious metal mining of the far West have for the most part swarmed out of the California hive, California ideas have not only been everywhere dominant over the field of industry, but have stemmed the tide of Federal land policy and given us a statute-book with English common law in force over half the land, and California common law ruling in the other.[38]

Nearly all of the new commonwealths passed laws which officially recognized the existence of codes and customs that had been established on California models, and several of the commonwealths tried to translate portions of the codes into statute law. Idaho and Oregon undertook to provide standard rules for lode claims.[39] Nevada attempted an even more ambitious step by enacting an elaborate law to govern both mining districts

---

[38] Public Lands Commission, "Report of the Public Lands Commission Created by the Act of March 3, 1879, Relating to Public Lands in the Western Portion of the United States," *House Exec. Doc.*, 46 Cong., 2 sess., no. 46 (February 25, 1880), p. xxxiv.

[39] Yale, *Legal Titles*, pp. 84-85.

and claims. The Nevada statute had to be repealed before it was a year old, because the provisions of the act were too complex and were coupled with an unprecedented attempt to make the miners pay "assessment dues." [40]

Arizona's territorial legislature, on the other hand, showed that it was possible to attain the goal of a single code for a whole state or territory. In 1864 it approved a measure that was "far more comprehensive than any legislation attempted in the United States upon mining rights" up to that time.[41] The *Mining and Scientific Press* described it as a "very excellent code," "far preferable to the multifarious and generally ill-digested local laws which have hitherto prevailed in California and Nevada." [42]

These state and territorial statutes represented both an improvement upon the parent model and a step toward the ultimate end, which was the establishment of a national mining code. In all the years since 1848 Congress had maintained a strict silence upon the subject of mining west of the Great Plains. The policy of withholding western mineral lands from sale or preëmption still prevailed, with the absurd result that the people of the United States were legally the owners of thousands of square miles of soil that had been occupied for years

[40] *Mining and Scientific Press*, May 5, August 11, 1866, February 23, 1867.
[41] Yale, *Legal Titles*, p. 85. But apparently the code did not include placer mining.
[42] *Mining and Scientific Press*, December 24, 1864.

under laws that were recognized by all of the states and territories beyond the plains. Annually millions of dollars' worth of gold and silver were being taken from the public domain for the private benefit of the men who were in actual possession.

To explain the paradox, the California judges and lawyers evolved the doctrine of "tacit consent," according to which Congress's persistent failure to provide rules for regulating mining was taken to mean that the federal government acquiesced in the measures taken by the miners in their own behalf. So long as the "tacit consent" continued, most of the miners were content to leave well enough alone.[43] They were satisfied to be occupiers rather than owners.

There were a few thoughtful persons who argued that the miners would be less nomadic if they could buy title to the land they were working, and thereby acquire a permanent interest in both their claim and their community. This opinion was not shared by the majority, who were able to point out that the gold and silver industry would always be an unstable business that could last in any one area only so long as the mineral deposits continued to yield.[44]

---

[43] On tacit consent, see *Irwin* v. *Phillips,* as quoted above, p. 221; on miners' attitude, see *Sacramento Weekly Union,* February 19, 1853, January 15, 1859, and "Editor's Table," *Hutchings',* III, 383.

[44] The chief proponents of selling the lands to the miners were John S. Hittell and the successive editors of the *Sonora Herald.* Twice bills on this subject were introduced into the California legislature, only to be voted down.

With the appearance of types of mining that required a large capital investment, there was some reversal of sentiment. Mining men came to feel that the lack of an unquestionable title was a deterrent to the advancing of capital and credit. English investors were known to be especially reluctant to sink their funds into mines that were secured only by a right of occupancy.[45]

This circumstance alone would not have been sufficient to reconcile the West to a sudden abandonment of the federal policy of doing nothing. What caused western leaders to alter their views was a growing uneasiness over the intentions of the federal government. There were legal precedents which seemed to indicate that the United States could, if it so desired, regard the miners as mere trespassers upon the public domain, and therefore as subject to ejection at the will of the paramount owner.[46] In November 1858, the United States attorney general intervened in a lawsuit over a California quicksilver mine in a manner which suggested that these precedents could easily be transformed into active principles.[47] A few years later another attorney general urged the President to eject unauthorized persons from some

[45] George Gordon, *Mining Titles: Are There Any—What Are They? A Letter* (San Francisco, 1859); *Sacramento Daily Union*, January 28, 1856; *Sacramento Weekly Union*, December 13, 1856; *Mining and Scientific Press*, June 9, 1862; *Meadow Lake Morning Sun*, June 6, 1866.

[46] *U. S. v. Gear*, 3 Howard (44 U. S.) 120, especially 133 (1845); *Cotton v. U. S.*, 11 Howard (53 U. S.) 229 (1850).

[47] Cf. *Sacramento Weekly Union*, November 27, 1858; *North San Juan Hydraulic Press*, same date; San Francisco *Daily Alta California*, December 6, 1858.

coal mines that had recently been discovered in California.[48] Almost simultaneously the chief justice of the California supreme court was so carried away by the nationalistic fervor of the Civil War that he brusquely shattered most of the legal props upon which the doctrine of tacit consent had been resting.[49]

Close upon the heels of these shocks to western complacency came definite signs of a change in federal legislative policy. The secretary of the treasury suggested the sale of the mineral lands as a means of paying the heavy Civil War debts, and Congressman George Washington Julian of Indiana, chairman of the House Committee on Public Lands, began a campaign to have the mineral lands surveyed and sold.[50]

In the early months of 1866 two mineral land bills were before Congress: Julian's, in the House, and one introduced by John Sherman, of Ohio, in the Senate. Senators Stewart, of Nevada, and Conness, of California, decided that it was no longer safe to remain on the defensive. Some of their constituents were already expressing agitation over the alleged attempts of the Central Pacific Railroad to extend their land grants over mineral-bearing soil, while the appearance of the Julian

[48] Edward P. Weeks, *A Commentary on the Mining Legislation of Congress, with a Preliminary Review of the Repealed Sections of the Mining Act of 1866* (San Francisco, 1877), pp. 1-2, fn.

[49] Wiel, *Water Rights*, I, 93-94.

[50] George W. Julian, *Political Recollections, 1840 to 1872* (Chicago, 1884), pp. 284-286; *Congressional Globe*, 38 Cong., 2 sess., pp. 562, 587, 684-687.

and Sherman bills was sufficient to produce in California a miners' convention that went on record as bitterly opposing any attempt to sell the lands upon which the mining industry depended.[51]

Stewart and Conness determined to carry the war into the enemy's camp. They undertook to rewrite the Sherman bill along such lines as would guarantee to the miners the continuance of all the privileges they then possessed, while at the same time protecting them from adverse governmental action in the future. In the opinion of Stewart and Conness, if the federal government was to venture into the field of mining legislation for the first time, then it must be upon the miners' rather than the government's terms.[52]

The act which became law in July 1866 was drawn in accordance with this point of view.[53] It established three fundamental principles:

First—That all the mineral lands of the public domain should be free and open to exploration and occupation;

Second—That rights which had been acquired in these lands under a system of local rules, with the apparent acquiescence and·sanction of the government, should be recognized and confirmed;

[51] The protest against both developments is reported in *Mining and Scientific Press*, November 11, December 9, 23, 1865, January 6, 13, 20, February 10, 1866.

[52] Cf. Julian, *Political Recollections*, pp. 287-288. Stewart was the real leader, though Conness was chairman of the Committee on Mines. For the debates, see *Congressional Globe*, 39 Cong., 1 sess., pp. 1844, 1865, 2851, 2888, 2957, 3126-3127, 3225-3237, 3451-3454.

[53] 14 U. S. Statutes at Large (1865-1867), c. 262 (pp. 251-253).

Third—That titles to at least certain classes of mineral deposits or lands containing them might be ultimately obtained.[54]

The first principle was a renunciation of the charge that the miners were trespassers. The second was a full recognition of the existing miners' codes and of the property rights based upon them, and it was also an assurance that the miners might continue their system of self-legislation in the future. The third was a decree that title to some of the mineral lands might now be obtained by those who were already in occupancy, provided they would pay the reasonable price of five dollars per acre plus the costs of surveying and recording. For the moment permission to secure title was confined to lode claims.

Four years later the provisions of the act were extended to placer mining, and in 1872 the lode and placer clauses were brought together into a single document, after alteration and improvement. In the Congress of 1873-74 the law was again changed slightly, in connection with the compiling of the *Revised Statutes of the United States*. The three fundamental principles of the act of 1866 were preserved in each of the successive revisions.[55] In view of the nature of the second principle,

---

[54] Curtis H. Lindley, *A Treatise on the American Law Relating to Mines and Mineral Lands within the Public Land States and Territories* (San Francisco, 1897), I, 61-62.

[55] 16 U. S. Statutes at Large (1869-1871), c. 235 (pp. 217-218), 17 U. S. Statutes at Large (1871-1873), c. 152 (pp. 91-96); Revised Statutes of the U. S. (1875), tit. xxxii, ch. 6 (pp. 426-434).

that meant that the United States incorporated into its permanent body of law the customs and usages which had first been adopted in California during the Gold Rush and subsequently had been developed and applied throughout "Greater California."

The *Mining and Scientific Press*, always the miners' spokesman, characterized the new law concisely when it said:

This bill [of 1866] is nothing more or less than putting into the form of a congressional act those local laws, which a practical experience of sixteen years has induced the miners of the Pacific Slope to adopt as the "Rules and Regulations" by which they have almost unanimously agreed among themselves to be governed, in the absence of all legislative enactments.[56]

The influence of the federal statutes made itself felt only slowly. The first attempts to secure patents bogged down in a precedentless morass, and the expense proved to be unexpectedly high. It was said that the average total charge for a patent was one thousand dollars, and that the holders of claims of dubious value were hardly justified in making so large an outlay.[57] Under the terms of the acts, obtaining a title was permissive rather than mandatory. Any miner who so desired could continue to operate under the local code of his district. With no immediate compulsion upon them to patent, and with

[56] *Mining and Scientific Press*, July 14, 1866.
[57] San Francisco *Daily Evening Bulletin*, January 13, 1869, January 11, 1871; *Mining and Scientific Press*, September 18, 1869.

high costs and legal intricacies to be faced, the miners were slow to take advantage of the new laws.

Number of Claims Patented in California[58]

Year ...	1867	1868	1869	1870	1871	1872	1873
Lode ...	4	2	6	6	13	54	84
Placer ..	1*	..	..	2	36	79	130

* This patent, three years before placer patents were legal, is not explained in the Public Lands Commission's tables.

At the time when the original act of 1866 was before Congress, Senator McDougall of California predicted that posterity would nickname it "a bill to promote litigation, create controversy, and occasion difficulties." [59] Fourteen years later the Public Lands Commission was about ready to acknowledge that McDougall had been right. After a survey of all the American mineral regions, the Commission concluded:

We find an extraordinary and characteristic difference between the mineral development east of the Missouri and that west. The first is almost absolutely exempt from litigation growing out of conditions of government conveyance [of title]. The other is a history of the most frequent, vexatious, costly, and damaging litigation.[60]

The commission believed that there were two main causes of the excessive proportion of western lawsuits.

[58] Thomas Donaldson, *The Public Domain, Its History, with Statistics, with References . . . Public Land Commission. Committee on Codification* (3rd ed.; Washington, 1884), pp. 325-327.
[59] *Congressional Globe*, 39 Cong., 1 sess., p. 3236.
[60] Public Lands Commission, "Report," *House Exec. Doc.*, 46 Cong., 2 sess., no. 46, p. xxxv.

The first was "the recognition by the law of the local customs and regulations." This recognition, the commission said, meant that the United States permitted the local codes to determine the conditions for locating, marking, and recording claims. This, in turn, opened the way for endless chicanery and carelessness. When the codes were framed a small group of miners were able to twist the local rules to suit their purpose, and thereafter the effectiveness of the rules was largely dependent upon how faithfully and how honestly the record of claims was kept by a recorder who was neither bonded nor under oath.[61]

Congress was not unaware of the weakness at the foundations of the law. In the revision of 1872 Congress specified for the first time what information the district records were to contain and what amount of work was to be necessary in order to hold a claim, while it also inserted a requirement that claim boundaries were to be clearly marked.[62] Congress did not, however, provide any safeguards against the loss or mutilation of the records, even though upon those records depended the outcome of lawsuits involving millions of dollars. Colorado had already suggested a simple way to avoid trouble. In an act of 1861 its territorial legislature wisely decreed that the records, laws, and proceedings of each lode dis-

[61] *Ibid.*
[62] 17 U. S. Statutes at Large (1871-1873), c. 152, sec. 5 (p. 92); repeated in Rev. Statutes, sec. 2324 (p. 428).

trict were to be filed in the office of the county clerk.[63]

The second cause of lawsuits, in the opinion of the Public Lands Commission, was the standard western practice in regard to lodes, or veins. Under both the English common law and the Spanish ordinances, the vertical extensions of the surface boundaries of a claim were the underground limits of what the owner might mine. Disregarding precedent, the early Californians decided that the vein was the principal thing, and that the claimant to the upper part of a vein should have the right to follow it downward to an indefinite depth and in any direction, even if that meant tunneling under someone else's surface property. They declared further that the claimant should have the rights to all "dips, spurs, angles, and variations" of his main vein, regardless of surface boundaries.[64]

From California this practice spread to all parts of the West and hence was written into the federal statute of 1866. Its inclusion in the first congressional enactment was made the more certain by the fact that fourteen years earlier Senator Stewart had been the chairman of a committee which had inserted it into the much copied

[63] Yale, *Legal Titles*, p. 85. In the early fifties Mariposa and Nevada counties, California, both made provision for recording claims at the County Clerk's office. San Francisco *Daily Alta California*, June 30, 1851; Grass Valley *Daily National*, January 30, February 16, 1865.

[64] Lindley, *Treatise*, I, 49. The lead miners of Derbyshire, England, had a similar provision. The provision appears in the Tuolumne County convention's rules of 1851 (San Francisco *Daily Alta California*, July 11, 1851) but must have been in use before that time.

Nevada County laws of 1852.[65] The original intent of the provision was to protect the quartz miner from the financial loss that would be incurred if he erected costly machinery on a given tract, only to discover that no more than the tail end of the vein was included within his vertical boundaries, while the rest of the ribbon of ore slanted under his neighbor's land.

In practice, the geological and legal difficulty of identifying and charting a particular vein on its downward course was so great that in almost every thickly settled lode area lawsuits between adjacent companies were a standard occurrence. When the act of 1866 came before the federal and the several state supreme courts, the judges set to work to whittle down the unlimited license implied in the vein provision. Congress retaliated by passing the much discussed "apex law" of 1872.[66]

This latter statute provided that the miner must locate his surface boundaries, which must form a rectangle, in such a way as to include the top, or "apex," of the lode that he was claiming. If he misjudged the course of the lode, and hence missed the apex, then he lost the right to follow the vein, and in no case was he to be allowed to pursue the vein beyond the vertical extension of the ends of his rectangular claim. If he guessed correctly

[65] William M. Stewart, *Reminiscences of Senator William M. Stewart of Nevada*, ed. by George R. Brown (New York and Washington, 1908), p. 127.
[66] 17 U. S. Statutes at Large (1871-1873), c. 152, sec. 3 (pp. 91-92). The background and subsequent history of the law are discussed in Lindley, *Treatise*, I, 64-77.

the location of the apex, then he could trace the vein for an indefinite distance through the vertical extension of the sides of his claim. This was essentially a compromise between the unbounded California precedent and the traditional European practice. The provision for differentiating between the end and side limits was understandable, but the theory of the "apex" was an artificial concept original with Congress. It introduced a new speculative element into quartz claiming and proved a prolific cause of litigation.

Because of the lode law and the inherent weaknesses of the records and rules of the local districts, California's legal legacy to the nation has not been an unmixed blessing. In their own day—1848, 1849, and the early fifties —the Argonauts were legislating as wisely as they knew, and in view of the social difficulties of the time, and of the American ignorance of mining law, it must be admitted that they achieved quite an impressive degree of success. As the fifties advanced, several serious flaws became apparent, and some attempt was made to correct them by local enactment.

Then came the sudden necessity of applying the still imperfect codes and customs to a vast new area beyond California's boundaries. It was then that the worst abuses appeared. In several of the new states and territories the financial stakes were much higher than they had been in California, and the incentive to litigation was correspondingly greater, while in others novel geological con-

ditions raised problems not fully anticipated in the precedents inherited from the Golden State. The Comstock
Lode, in particular, was far more productive of lawsuits
than California had been.[67] At that point in the development of American mining law, the federal and state or
territorial legislatures should have entered the field with
comprehensive and careful provisions. They, however,
felt hampered by the completeness with which the
whole life of the mining west was permeated by the
"California common law." In the words of Chief Justice
Chase of the United States Supreme Court:

> We cannot shut our eyes to the public history, which in-
> forms us that under this [local] legislation, and not only with-
> out interference by the national government, but under its
> implied sanction, vast mining interests have grown up, employ-
> ing many millions of capital, and contributing largely to the
> prosperity and improvement of the whole country.[68]

One may well ask whether Congress's action could
not have been to confirm property rights that already
existed, and then, with due deference to provisions that
had been shown to work well, to set up for the future a
true national mining code.

[67] *Mining and Scientific Press*, September 14, 28, 1863, November 12,
1864; Lord, *Comstock Mining*, pp. 133-173. Cf. report of Nevada state
senate, in Browne, *Report* (1867), pp. 227-229.
[68] *Sparrow* v. *Strong*, 3 Wallace (70 U. S.) 97, 104 (1865).

XIV

THE SURVIVAL OF THE FITTEST
1860-1873

By the opening of the sixties, California mining was entering its mature years. Behind it lay the days of easily worked, shallow deposits. Behind it lay the times during which tens of thousands of miners, laboring at a thousand different camps, were giving the state a huge annual yield. The last "big" season was that of 1856, when the output was $57,500,000. During the three seasons that were left in the fifties the gold product fluctuated between $43,600,000 and $46,590,000. Then came the sixties and the final decline.

The fundamental cause of this decrease was the simple fact of the impaired richness of the deposits themselves. The best of the accessible surface diggings "played out" during the early fifties. The river beds made their last great contribution in 1856 and 1857. Thereafter California had to rely primarily upon the deep diggings and quartz veins.

The successively smaller yields during the first half of the sixties are to be regarded as representing the transition from a productive system based upon several types of mining to a system based chiefly upon two. When the total annual amount finally fell below eighteen mil-

240

lion dollars, in 1865, it had reached its permanent annual average. The figure attained in that year was maintained with remarkable fidelity for the next two decades. Until the seasons of 1883 and 1884 the yearly output was always between fifteen and twenty million dollars, and the departure from standard at that time was the result of certain legal prohibitions rather than of a sudden change in the mines themselves.

Fiscal Year	Gold Produced in California[1]
1860	$44,095,163
1861	41,884,995
1862	38,854,666
1863	23,501,736
1864	24,071,423
1865	17,930,858
1866	17,123,867
1867	18,265,452
1868	17,555,867
1869	18,229,044
1870	17,458,133
1871	17,477,885
1872	15,482,194
1873	15,019,210

In all probability the transition from the high averages of the fifties to this permanent level would have been

[1] Hill, "Historical Summary," U. S. Bureau of Mines, *Economic Paper*, no. 3, p. 20.

spaced over the whole decade of the sixties if it had not been for the acceleration caused by two factors. The first and more important was the exodus to Nevada and the new mining frontier. How many miners departed who would otherwise have remained in the Golden State it is impossible to say. Doubtless many would soon have abandoned California mining in any case, because of its declining appeal to the man of small capital.

The second factor was the occurrence of three unusual seasons in succession. The season of 1861-62 brought the worst floods in California's history. Those of 1862-63 and 1863-64 together constituted the most prolonged and severe drought the state has known.[2] Both the excessive supply and the excessive lack of water caused great damage to the mineral industry, and at a time when it was least able to bear injury.

These two accelerating factors combined with the falling off in the richness of the diggings to cause a sharp decline in the size of the population of the mineral region. Nearly all of the mining counties reported a smaller population in 1870 than they had in 1860. El Dorado, Tuolumne, and Calaveras lost almost half their citizens during the decennial period, while Sierra lost more than half. Only three bona fide mining counties reported an increase, and in two of the three the gain was

[2] McAdie, "Rainfall of California," *University of California Publications in Geography*, I, 212.

very slight. The rest suffered losses that ranged from a few hundred to a few thousand.[3]

A large part of this decline was caused by depletion of the ranks of working miners. According to John S. Hittell's estimates, the number of miners in California was one hundred thousand in 1861, but less than thirty thousand in 1873. Hittell further believed that three-fifths of the total in 1873 was composed of the humble Chinese, who patiently devoted themselves to working and reworking placers that had been abandoned by white men.[4]

It might be supposed that the presence of so many Chinese would have given the mining counties an even more cosmopolitan character in 1870 than they had had in 1860. The contrary was true. Almost all of the mining counties reported a larger proportion of Americans in 1870 than they had returned ten years earlier. Behind the apparent paradox was California's change from a predominantly mining state to one in which agriculture, stock raising, lumbering, and commerce were the growing interests. Hittell, out of his unrivaled knowledge of the state's economic development, revealed the altered situation with his remark that "mining was until about

[3] Census figures in this chapter from *Ninth Census, 1870*, I (*Population*); 14-16, 304.
[4] Hittell, *Mining in the Pacific States*, pp. 20-21; San Francisco *Daily Alta California*, February 25, 1873, reprinted in Hittell's *Resources of California*, pp. 296-298.

1860 the chief industry of the State, but it has now been surpassed by both agriculture and manufactures." [5]

An important phase of this reversal in relative significance was the partial replacement of mining by the other arts in many of the foothill counties. During the sixties a considerable number of small farmers, viniculturists, fruit growers, and dairymen "took up" homesteads in the foothills, while other men grazed sheep and cattle on the unfenced range along the Sierras' flanks.[6]

The advent of rural husbandry changed greatly the racial composition of the mining counties, for the agricultural population was much more "American" than were the miners. It was significant that in the census of 1870 the greatest increases in the proportion of "native-born" were in those former mining counties in which agriculture had made the greatest progress.

In 1860 the foreign-born had been in the majority in all three of the main counties of the Southern Mines, in two of the northwestern counties, and in one of the central counties. In 1870 the foreign-born were dominant in only one southern and two northwestern counties, although their numbers came within a few hundred of equaling those of the native-born in the other two

[5] Hittell, *Resources of California*, p. 296.
[6] Wilson Flint, "Textile Fibres of the Pacific States," California State Agricultural Society, *Transactions during the Years 1864 and 1865* (Sacramento, 1866), pp. 284-285; *Pacific Rural Press*, November 4, 1871; North San Juan *Hydraulic Press*, May 19, 1860; *Sacramento Weekly Union*, February 26, June 25, October 8, 1859, February 18, 1860, February 9, March 16, 1861.

southern and in one central county. The three counties in which the foreign-born were still supreme in 1870 were ones whose economic and social condition entitled them to the distinction of being the most backward parts of the mineral region.

More important than the racial change was the decline in the amount of gold produced and the number engaged in producing it. The most serious losses were in the Southern Mines. There, by the later sixties, the majority of the towns, being dependent on placer diggings that no longer gave satisfactory returns, were falling into decay. At most of them a limited amount of mining was still carried on, especially during the season of the year when water was most abundant, but the number of paying claims and working miners was small indeed when compared with the vanished heyday of the early and middle fifties.[7]

Even the once fabulously rich Columbia Basin felt as early as 1861 the chilling touch of that paralysis which turns booming camps into crumbling ghost towns. By 1867 the basin's queen "city" of Columbia, formerly the proud "Gem of the Southern Mines," was described as being in a state of "almost total desertion," with the few

[7] *Stockton Daily Independent*, April 27, 1868; *Columbia Citizen*, August 4, 11, 1866, February 9, April 6, 27, May 4, 1867; *Mining and Scientific Press*, September 14, November 23, 1867; San Francisco *Daily Evening Bulletin*, January 12, 1865; Lang, *Tuolumne County*, pp. 227-228, 233-235; Browne, *Report* (1868), pp. 13, 35, 50; Rossiter W. Raymond, *Statistics of Mines and Mining in the States and Territories West of the Rocky Mountains* (Washington, 1870), pp. 24, 29.

remaining miners already undertaking to sluice away the "city lots" in their efforts to find unworked ground.[8]

The severity of the decline in the section as a whole was revealed by the comparative census figures for the three counties which together embraced the greater part of the auriferous ground in the Southern Mines:

County	1860	1870
Calaveras	16,299	8,895
Tuolumne	16,229	8,150
Mariposa	6,243	4,572

Deprived of much of their former population, the Southern Mines suffered a still further reduction in the relative importance of their gold product, as compared with that of their northern neighbors. According to Wells, Fargo & Co., which transported most of the treasure of the western states, the records of gold received at San Francisco showed that from 1861 through 1865 the gold shipments from the Southern Mines were never more than 26 per cent of the state's total, and sometimes were only 20 per cent.[9] Similarly, in 1866 a competent mining expert expressed the belief that about 80 per cent of the gold then being produced in California was derived from deposits located in the central and northwestern sections.[10]

[8] *Mining and Scientific Press*, November 23, 1861, March 16, 1867; and cf. *ibid.*, April 18, 1862, and Browne, *Report* (1868), p. 36.
[9] Browne, *Report* (1867), p. 51.
[10] *Ibid.*, p. 40; estimate by William Ashburner.

It was expected by many that the development of the quartz industry would compensate for the deficiencies of the southern placers. In certain restricted areas, and for brief periods of time, this did indeed prove true, but in the Southern Mines generally, as in the remainder of the state, quartz proved an elusive hope that did not live up to the anticipations of its advocates.

As between the two sections of the Northern Mines, the northwestern was still much the less important. In fundamental conditions it had changed little; in population it had not grown appreciably. One northwestern mining county was able to report a net gain of twenty-nine persons between 1860 and 1870, and one county that was primarily devoted to agriculture and commerce more than doubled its slim population, but for the most part the section continued to be a sparsely settled, retarded mining frontier.[11]

During the sixties there was some improvement in communications between the northwest and the outer world. More significantly, in 1871 and 1872 the railroad thrust its way up to the southern limits of the section.[12] But even with greater accessibility the northwest was still regarded as an area to be recommended chiefly "for such miners as are fitted to endure the hardships of a

[11] *Mining and Scientific Press*, July 18, 25, 1868, December 7, 1872, January 25, 1873; San Francisco *Daily Alta California*, July 10, 1872; Hittell, *Resources of California*, pp. 83-85.

[12] Cronise, *Natural Wealth*, p. 568; "From Trail to Rail—The Story of the Beginning of Southern Pacific," *Southern Pacific Bulletin*, XVIII (March 1930), 14.

rough and laborious life," since "the country is rugged, the climate wet and cold, the roads bad, and there is some danger of Indians." [13] On the other hand, precisely because it had always been backward, the northwest could still offer so large an amount of "comparatively undeveloped" auriferous ground that it was "the best region in the State for the miner who wants to work on his own account, and on a small scale." [14]

In the central section conditions were more varied. Two of the central counties lost about half their inhabitants during the decade, one lost a substantial proportion, four very nearly stood still, and one made a real advance. All would have suffered more severely than they did if farmers, dairymen, and stock raisers had not come to live in their valleys.[15]

The hardest hit of any was Sierra County, which declined from a population of 11,387 to 5,619—more than 50 per cent. An unfriendly natural environment was the cause of most of its trouble. As the name suggests, the county was located far up in the mountains, contiguous to the state of Nevada. The lowest point in the county was said to be 2,000 feet above sea level, while most of the mining camps were situated at 4,500 feet, or higher. Most of the travel and freighting had to be by mule-

[13] First quotation from Cronise, *Natural Wealth*, p. 568; second from Browne, *Report* (1867), p. 66.

[14] Browne, *Report* (1867), p. 66.

[15] Browne, *Report* (1868), pp. 71, 82, 92, 157; Cronise, *Natural Wealth*, pp. 226, 229, 234, 244.

back, and for several months each year deep drifts of snow left the camps almost in isolation.[16]

As long as it was possible to work Sierra's placers with the primitive rocker and long tom, and with short tunnels and shafts, the county was prosperous. When that ceased to be feasible, the county's natural disadvantages began to tell against it. Sierra possessed some of the best deep gravel deposits in the state, and one of the most profitable quartz mines. So great, however, were the topographic obstacles to developing its water resources that many of its deep gravel claims had to be worked through tunnels rather than by hydraulicking, and where hydraulicking was used, it was often restricted to the four or five months during which nature provided the proper conditions. In quartz, the quality of the county's paying mines was high, but their number was too few to make a major contribution to the local economy. Presumably the difficulty of importing adequate machinery was a factor in hindering the opening of additional mines.[17]

El Dorado County, whose loss of a little less than 50 per cent placed it next after Sierra, was the antithesis of that county. As the site of the original gold discovery and the home of river mining, it had the largest population of any mining county in both 1850 and 1860. It

[16] Browne, *Report* (1868), pp. 137-138.
[17] Browne, *Report* (1868), pp. 138, 144-148; Cronise, *Natural Wealth*, pp. 230-231, 582-583; *Mining and Scientific Press*, June 1, 1861, November 12, 1870.

was the first mining county to be tapped by a railroad; through it ran the most traveled of the several wagon roads that crossed the Sierras to Nevada; agriculture made greater progress in it than in any other mountain county; it had an important lumber industry; for its mines it had an extensive system of water ditches.[18]

Even with all these advantages and all these sources of additional population, El Dorado was unable to hold its people during the sixties, for it had become subject to the same disease that was afflicting the Southern Mines: its mineral deposits were no longer of especial significance. The "gulch and bar diggings" had been "pretty nearly worked out." The river-bed claims had ceased to pay. Hydraulicking was confined to "small detached patches" of "auriferous detritus." Quartz mining was "not extensively carried on." Few of its gravels and veins impressed contemporaries as being first-rate. Meanwhile, Nevada territory was beckoning from across the Sierras. The result was what the census figures revealed.[19]

A different set of circumstances determined the fate of Yuba County. It, too, had great advantages in regard to transportation, thanks to the Feather River. During the later fifties and the sixties it, like El Dorado, came to number many farmers, stock raisers, and lumbermen among its citizens, and it included within its boundaries

[18] Browne, *Report* (1868), pp. 81-82; Cronise, *Natural Wealth*, pp. 246-250.

[19] Quotation on gulch and bar from Cronise, *Natural Wealth*, p. 249; on hydraulicking and quartz from Phillips, *Mining and Metallurgy*, p. 58.

Marysville, once the third city in the state. To serve its mines it had one great water ditch and numerous small ones. Nevertheless, it failed to maintain the size of its population during the decade.[20]

Why? Because its easily worked placers and river-bed claims "played out" at the end of the fifties and left behind them a veritable colony of dying camps. One of the camps was rejuvenated in 1863-64 when Yuba began to develop its only quartz district, but that was hardly enough to compensate for the previous losses. The county's real mineral wealth in the later era was a rich belt of deep gravels that crossed the upper part of the county. Although outstanding success attended the working of these gravels, the Census showed that Yuba's mining population had declined sharply between 1860 and 1870. This was no paradox. It meant, simply, that Yuba's wealth was being exploited by means of the hydraulic, which was preëminently a labor-saving method of mining. For the thousands who had once used the rocker and long tom there was no longer any place.[21]

Placer County's decline can be explained in the same fashion. There the advance of California civilization did everything possible to bolster up the county. In addition to gaining a considerable number of farmers and lumbermen during the sixties, Placer was chosen as the route

[20] Cronise, *Natural Wealth*, pp. 300-303. The Feather River's value to commerce declined as it began to silt up with hydraulic debris.

[21] Browne, *Report* (1868), pp. 13, 148-155; Cronise, *Natural Wealth*, p. 302; *Grass Valley National*, March 8, 1864.

for the Central Pacific Railroad. The latter was built "through to the summit" of the Sierras in 1867—a fact which meant that Placer acquired three railroad towns and all the attendant business that implied. Nevertheless, Placer's population in 1870 was smaller by nearly two thousand than it had been in 1860.[22]

The explanation was that during the sixties Placer's surface diggings produced little and its quartz mines were inconsiderable. What Placer did have was large tracts of exceedingly rich deep gravels. These were extensively and profitably worked, sometimes by tunnels, more frequently by hydraulicking and its supplement, tail sluicing. The resulting product of gold was large, but there was not enough employment, especially of an independent, small-capital kind, to hold the miners within the county.[23]

Finally, there was the case of Amador County. There the wonder was not that its population declined by something more than a thousand, but rather that it managed to maintain its numbers as well as it did. Amador was one of the smallest of the mining counties. No railroad served it and its attempt to build a highway across the Sierras brought little benefit. While it had a limited amount of agriculture and stock raising, during

[22] Cronise, *Natural Wealth*, p. 244; Browne, *Report* (1868), p. 92; *Sacramento Daily Union*, November 30, 1867.
[23] Browne, *Report* (1868), pp. 92-108; Phillips, *Mining and Metallurgy*, pp. 58-59; *Mining and Scientific Press*, January 11, 1868, September 3, 24, 1870.

the sixties both were in dubious condition. After the fifties it had few important surface placers and no extensive hydraulic claims.[24]

What saved the county was its quartz mines—or, rather, its quartz mines and the ability and capital that were brought to bear upon them. Quartz mining began in Amador in 1851, but the greater part of the fifties was passed in learning how to master that most difficult of all types of gold working. Many of the operators became bankrupt and others withdrew in discouragement. Still the persevering few had the faith to continue.[25] The most notable of these was Alvinza Hayward. He began developing his claim in 1853, in company with several partners. At the end of four years all were "dead broke," and all but Hayward abandoned the project. Hayward borrowed from nearby friends and obtained credit from local tradesmen. When he had sunk his shaft down to a depth of four hundred feet, he finally struck "pay ore." From that time forward the mine yielded well. By 1872 Hayward was reputedly the richest man in California.[26]

Hayward's success, and the similar triumphs scored by

[24] Jesse D. Mason, *History of Amador County, California, with Illustrations and Biographical Sketches of its Prominent Men and Pioneers* (Oakland, California, 1881), pp. 98-99, 110; Browne, *Report* (1868), p. 71.

Of the two central counties not discussed here, Butte's history resembled that of Yuba and Plumas' that of Sierra.

[25] Mason, *Amador County*, pp. 110, 145; *Sacramento Weekly Union*, February 26, 1859.

[26] De Groot, "Mother Lode," *Overland*, 1st series, IX, 408; Cronise, *Natural Wealth*, p. 254.

several other Amador lode mines at the same period, convinced both mining men and investors that the ten mile piece of the Mother Lode included within Amador County was indeed "the most productive part of the [Mother Lode] belt," as a modern geologist has phrased it.[27] Able mining men and monied San Franciscans and Sacramentans—such as Leland Stanford—became interested in Amador mines during the latter half of the sixties.[28]

By the start of the seventies, several of the mines were paying richly and others were being developed. Collectively they were giving employment directly to a thousand men and indirectly to many more. Because of their constant need for heavy equipment and supplies, the quartz mines supported an iron foundry, lumber yards, and enough other establishments to give the leading town, Sutter Creek, the appearance of a New England manufacturing community. The presence of the dependent industries and the continuing profitableness of the mines gave Sutter Creek a life that was unusually stable for a mining town.[29]

Despite the impetus from its one big industry, Amador failed to maintain the size of its population during the decade, and one feels compelled to ask the question: just

[27] Knopf, "Mother Lode," U.S.G.S., *Professional Paper*, no. 157, p. 1.
[28] *Mining and Scientific Press*, October 10, 1868; Cronise, *Natural Wealth*, p. 254.
[29] Mason, *Amador County*, pp. 110-111; *Mining and Scientific Press*, June 11, 1864, February 4, 1871, July 20, 1872, May 21, 1910, June 24, 1916; *Sacramento Weekly Union*, February 26, 1859.

what did a mining county have to possess and to do in order to show the usual healthy advance in numbers that characterized most western American communities of the nineteenth century? The answer is provided by the history of Nevada County.

Nevada was the only mining county to make a significant gain in population. It grew from 16,446 in 1860 to 19,134 in 1870. Like many another county, Nevada had begun its life in 1848-1850 with a boom based upon shallow placers and river mining. When that collapsed, Nevada turned to the more permanent wealth of her deep gravels and quartz veins.[30]

In both fields severe difficulties had to be surmounted during the early years, yet in placer mining improvement seems to have been almost continuous after 1849, and in quartz mining steps forward were made steadily after 1853. Throughout the greater part of the fifties Nevada was thriving. Then came the early sixties. In spite of her prosperity Nevada shared with the rest of the state the ill effects of the "Washoe fever," the flood, and the drought. One observer claimed that the county's population declined by nearly one-third before the start of 1864. In the spring of the latter year the county began to show signs of recovery, and by August it was on the road back to vigorous economic health. Thereafter the flow of population was one of immigration, not emigration, as the county won back many of the miners

[30] *Bean's History*, pp. 11-12, 30-31.

who had left it a short time before for Nevada territory.[31]

This quick and effective recovery, contrasting so sharply with the condition of other parts of the California mines, was made possible by the unusual richness of the county's natural endowment, and the equally remarkable ingenuity and ability that was shown in the development of those resources. In minerals no county could surpass and few could equal the wealth of either Nevada's deep gravels or its quartz veins.[32] For its water supply it had in 1860 the best system of ditches in California, and the passing of the decade brought several extensions to this already excellent network.[33]

For transportation purposes its steeply hilly topography was hardly what one might have asked for, but the degree of ruggedness was not sufficient to prevent the early development of Nevada City into a trade and travel center of more than local importance, nor was it sufficient to prohibit the establishment of a large staging business through Nevada County to Washoe.[34]

Attracted by the potentialities of the county, a suc-

[31] Estimate of population loss from Rolfe, "Mines and Mining," *Bean's History*, p. 50. Other details from: Browne, *Report* (1868), pp. 118-129; *Grass Valley National*, September 8, 24, October 17, 1863, February 18, March 8, April 14, August 23, 1864; *Mining and Scientific Press*, August 27, 1864.

[32] Browne, *Report* (1868), pp. 111-137.

[33] Rolfe, "Mines and Mining," pp. 65-72.

[34] *Sacramento Transcript*, October 19, 1850; *Sacramento Daily Union*, July 26, 1855; *Grass Valley National*, October 6, 1863.

cession of men of outstanding ability and inventiveness drifted into Nevada. In placer mining they made their presence felt in 1850, and thereafter brought forward one improvement after another. Nevada was the first county to make use of water ditches. It claimed that it was the first county in which the long tom and the sluice were employed. It was the scene of the discovery of the principle of the hydraulic hose and of most of the subsequent improvements in that process.[35]

In quartz mining Grass Valley, Nevada County, early became, and has ever since remained, not only the most important center of auriferous vein mining in California, but one of the best known in the world.[36] It was in Grass Valley that much of the trial and error experimental work was done during the fifties.

Like Sutter Creek, Grass Valley showed the beneficial influence of the greater permanence of quartz as contrasted with placer mining, and of the fact that a comparatively large population could find employment in the mines, mills, iron foundries, machine shops, and other dependent industries. Visitors always commented upon the town's stable and well settled appearance. They often praised its "pretty cottages," its churches, and its

[35] *Sacramento Daily Union*, July 26, 1855; *Scientific Press*, October 8, 1870; *Bean's History*, p. 32. But on first use of the long tom, see *Sacramento Daily Union*, January 15, 1855.
[36] Phillips, *Mining and Metallurgy*, p. 60, *passim* (and note that this is an English book); also Browne, *Report* (1867), p. 52.

schools, while the local press never ceased to boast of the unusually large number of ladies and children.[37]

Grass Valley was also the only mountain town that could set up a strong claim to having developed an intellectual interest in the occupation in which most of its citizens were engaged. The first miners' discussion club in the state was established at Grass Valley in 1855, in order to facilitate the exchange of ideas relating to lode mines.[38] A few months later Warren B. Ewer began publishing in Grass Valley the first mining journal in the West. Ewer, a New Englander by birth and education, had once been the superintendent of a nearby quartz mine and was at that time the editor of the little local weekly newspaper. After two years his journal died of financial malnutrition, but four years later Ewer moved down to San Francisco to become the editor and publisher of the first successful mining publication west of the Rockies.[39]

Perhaps the explanation for this intellectual activity was, in part, the peculiar nature of the gold-quartz veins at Grass Valley. The Grass Valley veins were very rich, but also very narrow. The latter circumstance meant

[37] For visitors' comments, see *Sacramento Daily Union*, July 23, 1855; *Mining and Scientific Press*, July 6, 1861; for local press, see *Grass Valley Telegraph*, November 6, 1855; Grass Valley *Nevada National*, October 29, 1859, June 16, 1860.

[38] *Grass Valley Telegraph*, November 27, 1855.

[39] The first journal was *The California Mining Journal*, founded April 1856; the second was the *Mining and Scientific Press*, founded at San Francisco in 1860 by Julius Silversmith, taken over by Ewer in 1862.

that a large amount of barren surrounding rock had to be cut away in order to reach the thin ribbon of auriferous quartz. For that reason the cost of extracting gold was unusually high at Grass Valley, and it was "more than elsewhere in the Gold Region . . . necessary for the proprietors of the mines and mills to use every possible means to improve their machinery, so as to save the largest possible amount of the precious metal at the least expense." [40]

An interest in improved methods was thus forced upon the Grass Valley quartz men, and they responded by developing an inquisitive frame of mind that caused them to try out "almost every new-fangled idea," as the state geological survey put it, until they had evolved a paying technique for mining and milling. At the same time, in other respects than the width of the veins, geologic and underground conditions at Grass Valley were unusually favorable, so that there was ample justification for the attempt to overcome the handicap of the narrowness of the veins.[41]

The preëminence of Grass Valley in quartz and of Nevada County generally in placer mining was what the San Francisco press had in mind when it said: "Nevada is the leading mining county of California. It has the

[40] Geological Survey of California, *Geology*, Volume I: *Report of Progress and Synopsis of the Field-Work, from 1860 to 1864* (n.p., 1865), p. 290.

[41] William D. Johnston, Jr., "The Gold Quartz Veins of Grass Valley, California," U. S. Geological Survey, *Professional Paper*, no. 194 (Washington, 1940), pp. 1, 22-23.

largest mining population, the largest gold yield, the most thorough system of ditches, the most profitable quartz and hydraulic mines; and within its borders many of the most important mining inventions were made or first applied in this State." [42]

As with Amador, this high repute caused many veteran mining operators and a large amount of "outside" capital to come into Nevada in the later sixties and early seventies, after the bright glitter of Washoe's silver had lost some of its alluring qualities. When added to Nevada's own human and monetary resources, this made possible a vigorous development. In the latter half of the sixties, Grass Valley's quartz mines were producing $3,200,000 a year and were employing two thousand men, while the county's hydraulic claims and other placers were yielding about $3,500,000 and employing another two thousand men, several hundred of whom were Chinese. [43]

The history of this county showed that even the best of mineral areas was not proof against at least temporary retrogradation. It demonstrated that mining folk were so susceptible to "excitements" that they would leave even paying claims if they heard there was a new pot of gold at the end of the rainbow. Since Nevada had the maximum of factors in her favor, many of her wanderers

[42] *Mining and Scientific Press*, October 8, 1864, clipping San Francisco *Daily Alta California*.
[43] Quartz figures from Cronise, *Natural Wealth*, p. 580; placer figures from Browne, *Report* (1868), p. 122.

came back, and those who remained away were soon replaced by newcomers.

Elsewhere the losses from the mineral districts were permanent, because few counties were without their weak points. Some had exhausted the best of their mineral deposits, as was true of the Southern Mines generally and of El Dorado County. Others were held back by physiographic difficulties, as was the case in the northwest and in Sierra County. Still others suffered in spite of their wealth, because they were primarily dependent upon a labor-saving type of mining. That was the experience of Yuba and Placer counties.

This is the equivalent of saying that the mining of the later era was able to support only a very few of the districts and communities that it inherited from the times that had gone before. The Gold Rush had peopled the foothills. It had called into being commerce, transportation facilities, cities, and agriculture. It had brought civilization into a hilly or mountainous territory that otherwise would have been left for decades to the aborigines and wild beasts.

It was able to do so much because tens of thousands of men had believed that there they would quickly win their fortune. When the day of disillusionment came, the tens of thousands were speedily reduced to thousands. Some of the crowd stampeded off to new El Dorados. Others sought new occupations. The rest returned to their old homes. Behind them they left, like a courageous rear guard, a scattering of still vigorous camps whose claims still paid richly. But behind them they also left, and much more frequently, ghost towns and deserted villages that stood as lonely witnesses to this truth: mining is a good way to pioneer a territory, but a poor way to hold it.

XV

THE DISTANT PRIZE
1860-1873

The deterioration of the California gold region would have been much less severe if the miner and the investor had been willing to consult their own best interests. Scattered through the better districts were many good claims that needed only labor and funds to become paying mines.[1] That such claims went begging was mainly the result of the madly speculative frame of mind that dominated so many Californians during the hectic period that was ushered in by the discovery of the Comstock Lode.

It was not only that Californians found the whole region west of the Great Plains suddenly opening itself to exploitation; they found also that on the remote fringes of their own state there were gold and silver deposits whose existence no one had suspected hitherto. They found, moreover, that their amazing state contained copper, petroleum, coal, and quicksilver. As their newspapers cried out to them the rumors of one "great discovery" after another, the name "El Dorado" began to seem hopelessly inadequate for so wonderful a land. California had not merely gold, it had

[1] Cronise, *Natural Wealth*, pp. 570-572, 586-587.

every other mineral. To win his fortune one needed only to rush off to some unknown region, or to invest his savings in an untried project. The *Calaveras Chronicle* was not far from the truth when it remarked that "the only occupation in which men appear to engage without the least preparation or forethought, is mining." [2]

Most of the "excitements," whether over new deposits of gold or over some other material, proved false hopes that absorbed without recompense the time and money of those who participated in them. A few proved remunerative, but not to the persons who originally promoted them. With the benefit of hindsight it is easy to dismiss these diversions as side shows, and to recognize that gold mining in the best of the traditional California districts should have been adhered to as the main California performance. Contemporaneously this distinction was not always clear. The minor manias were allowed to draw both energy and funds away from the chief task.

The booms in silver and copper were good examples. Both metals, curiously enough, came into prominence as a result of the great events in Washoe. Of the two, silver had perhaps the greater long-term significance, since the search for it led to the pioneering of a difficult region that would otherwise have been long eschewed by settlers.

Nearly all of the important silver discoveries were in

[2] Mokelumne Hill *Weekly Calaveras Chronicle*, November 7, 1868.

the eastern part of the state. Some were high up on the eastern edge of the Sierras, such as those in Alpine County, which were located at altitudes of 5,000 to 7,000 feet, or the Cerro Gordo mines, in Inyo County, which were nearly 7,000 feet above the sea. The others were scattered through the arid tongue of land that thrusts its way up from the Mojave Desert between the Sierras on the one side and the Nevada boundary on the other.[3]

A more unpromising region could hardly have been selected for a new industry. Most of the region could be reached by wagons only if the drivers approached from the state of Nevada, or else made a long haul from Los Angeles across miles of desert, past the southern tip of the Sierras, and then up the long, rainless valley on the east side of the Sierras. At the end of such a journey the goods had to be packed up steep mountain trails in order to reach the mines themselves. In winter heavy snow rendered communications still more difficult at the higher altitudes, while at the lower levels continuously high temperatures made summer work of all kinds a severe strain upon men and beasts.[4]

There was the added difficulty that here, as in Nevada, the miners found that the treatment of silver was a much

[3] Partly shown on the map in *Mining and Scientific Press*, December 21, 1861.
[4] Browne, *Report* (1868), pp. 170-179; Hittell, *Resources of California*, pp. 9-11; Cronise, *Natural Wealth*, pp. 257-260, 280-288. Parts of Alpine and Mono counties could be tapped from western California.

more complex matter than the reduction of gold-quartz. They found further that many of their lodes were not of the highest grade, that others contained "rebellious" ores that were very hard to deal with, and that for most the costly process of smelting was necessary.[5]

Lack of capital was a further obstacle, and the scarcity of funds had the result that the majority of the operations were confined to mere prospecting and to limited efforts with inadequate machinery. At the close of the sixties the situation was improved in Alpine County by an influx of English and eastern capital, while at Cerro Gordo, Inyo County, the mines themselves began to provide much needed income. Nevertheless, it cannot be said that California ever became a great producer of silver. Most of the excitement over that metal occurred during the first half of the sixties, when the fame of Washoe was still the wonder of the world. Ten years later, all save a few of the silver mines were either entirely idle or producing only a small amount of bullion.[6]

Copper eventually suffered a similar collapse, although for quite different reasons. Copper was known to exist in California before the discovery of the Comstock

[5] *Mining and Scientific Press,* May 23, 1862, September 14, 1863, November 5, 1864, October 5, December 14, 1867, March 7, September 26, October 3, 1868, March 4, 1871.

[6] *Ibid.,* December 21, 1861, June 9, September 11, 1862, July 9, August 6, 1864, January 7, 1865, December 29, 1866, October 5, 1867, September 28, October 3, 1868, November 13, 1869, February 5, 1870; Hittell, *Resources of California,* pp. 330-331; Winchester, Scrapbook, pp. 54-62; Willie A. Chalfant, "Cerro Gordo," Historical Society of Southern California, *The Quarterly,* XXII (1940), 55-61.

Lode, but its potentialities were not appreciated until a Californian from Calaveras County caught the Washoe fever and crossed the Sierras. In Nevada he was impressed by the similarity of the Comstock silver ledges to some outcroppings near his former home. Returning to Calaveras County, he "located" what he thought to be a silver mine, but what proved to be a vein of copper. The customary California "excitement" immediately ensued.[7]

Once the value of cupriferous lodes was recognized, prospectors began establishing claims at many points in the Mother Lode counties, in Nevada, Yuba, and Placer counties, and in the northwest. The chief center of the new industry, however, remained in Calaveras County, near the site of the first discovery. There the appropriately named town of Copperopolis boomed into prominence during the first half of the sixties.[8]

In order to secure the capital to open their mines, the promoters imitated Washoe by incorporating, placing their stock on public sale, and courting the support of urban finance. During 1863, when the "fever" was at its height, there were 380 incorporations of copper companies in twelve months. San Francisco and Sacramento "capitalists" responded by investing in many of

[7] *Mining and Scientific Press*, July 6, 1861, August 17, 1863.
[8] *Ibid.*, July 6, 1861, December 6, 1862, January 19, August 17, 24, 1863, January 7, 21, 1865; M. M. O'Shaughnessy, "The Copper Resources of California," California Miners' Association, *California Mines and Minerals*, p. 205.

the new enterprises, and their unwise example was copied by men of small means, such as Henry George, the future author of *Progress and Poverty*.[9]

Aside from the general ignorance concerning the fine points of dealing with copper ores, the great handicap was the absence of adequate smelting facilities. Numerous attempts were made to establish smelters in California, but the most important undertaking failed, and the successful smelters were too limited in capacity and too dispersed in location to meet the need. This left the Californians with no alternative but to ship their ore to Boston, New York, or Swansea, England, and because of the high cost of transporting so bulky and heavy a cargo for so great a distance, they were automatically excluded from utilizing any save their richest ores, since average ores would not bring a high enough return per ton to compensate for the expense of transportation.[10]

Copper mining in California was thus a dubious affair from the start. At the time when it began, in 1860-61, conditions were temporarily favorable. The protective tariff duties of 1861 were having a beneficial effect upon the price of copper, and the ocean freights were at rea-

[9] *Mining and Scientific Press*, August 24, 1863, January 30, 1864; *Grass Valley National*, September 19, 1863; Henry George, Jr., *The Life of Henry George* (New York, 1900), pp. 138-141.

[10] *Mining and Scientific Press*, January 5, 12, March 16, July 20, August 24, September 7, 21, 1863, October 29, 1864, January 7, 21, 28, April 15, 22, 1865, January 13, February 17, 1866, January 5, 1867, January 18, 1868; Browne, *Report* (1868), pp. 207-211.

sonable figures. By the end of 1862 the California producers had succeeded in exporting from San Francisco $500,000 worth of ore, with Calaveras County contributing almost the entire amount. During 1863 and 1864 the exports were appraised at about $500,000 or $600,000 per year. In the next year both the quality and the quantity were increased, with the result that shipments for those twelve months reached the impressive total value of $1,500,000.[11]

Then disaster struck. Attracted by the high price, copper began coming into the world market from many different parts of the globe. Abruptly the price tumbled, and by mid-July 1866 it had ceased to be profitable to ship to England from San Francisco. At almost the same time ocean freights began rising, because California wheat was beginning to compete for tonnage. Caught between the two millstones, copper exports declined sharply.[12]

Having lost their market, most of the California producers discontinued operations during 1867.[13] A rise in the protective tariff duties brought a brief revival of hope at the end of the decade, but the competition offered by the expanding output of the great Calumet & Hecla mines of Michigan left little chance for California.

[11] *Mining and Scientific Press*, December 6, 1862, August 24, 1863, February 17, 1866.
[12] *Ibid.*, September 1, 1866, January 18, 1868.
[13] *Ibid.*, January 18, 1868.

The period closed with copper listed as one of California's several dead or dying mineral industries.[14]

The briefest of all the manias was the excitement over petroleum; that was born in the early months of 1865 and was dead a year later. In between its birth and death it succeeded in drawing money out of capitalists like Louis McLane and George T. Hearst (the father of the newspaper publisher) at the one extreme, and out of plain folk like young Frank Leach, a local printer, at the other. At the peak of the boom, near the close of 1865, about seventy-five oil companies were in the field, with the backing of $50,000,000 in "paper" capital.[15]

The most peculiar feature of this ill-starred excitement was that it centered in the undeveloped regions in the northwestern corner of the state, rather than in the present-day oil districts of southern California. A second peculiar feature was that none of the companies succeeded in "bringing in a gusher." Their wells produced a seepage of oil, but not a steady flow. What finally destroyed the prospects of the industry, however, was the inability of the San Francisco refiners to compete

[14] *Ibid.*, July 20, October 26, 1872; James Douglas, "Copper through Fifty Years," *ibid.*, May 21, 1910. From 1845 to 1875 Michigan produced 82.02 per cent of the copper produced in the U. S. California ranked second with 4.36 per cent. Carl E. Julihn, "Summarized Data of Copper Production," U. S. Bureau of Mines, *Economic Paper*, no. 1 (Washington, 1928), p. 17.

[15] *Mining and Scientific Press*, April 1, December 2, 1865, September 29, 1866; Frank A. Leach, *Recollections of a Newspaperman, A Record of Life and Events in California* (San Francisco, 1917), pp. 91-100.

FEEDING THE MILL.

HYDRAULIC MINING, WASHING DOWN BANK.

with the large quantities of cheap oil that were imported from the East as soon as a market for that item appeared in California.[16]

Still another mineral that Californians sought to exploit during the later era was coal. Because of the state's dependence on river and coastwise steam navigation, the fuel question was important, and it was early recognized that California's scanty supply of firewood was not an adequate answer. From time to time it was announced that important coal deposits had been discovered or were being developed, but, in actuality, most of them were poor in quality, or small in quantity, or inaccessible in location. Almost all of the California coals were inferior brown lignites that were of limited value for industrial and household purposes.[17]

Despite these drawbacks, one noteworthy coal district did come into being near Mount Diablo, on the peninsula which lies between San Francisco Bay and the San Joaquin River. Operations there began at the start of the sixties and progressed fairly rapidly. In 1863 the mines were able to produce 37,000 tons during the twelve months, and a decade later 180,000 tons during a similar period.[18]

[16] *Mining and Scientific Press*, October 21, December 2, 1865, January 18, 1868; Cronise, *Natural Wealth*, p. 625.

[17] Watson A. Goodyear, *The Coal Mines of the Western Coast of the United States* (San Francisco, 1877), pp. 5-6; San Francisco *Daily Alta California*, December 1, 1872; *Mining and Scientific Press*, February 8, 1873.

[18] *Mining and Scientific Press*, January 16, 1864, February 8, 1873.

The mines were giving employment directly to about five hundred men in 1869, and were the chief support of a half-dozen small towns in Contra Costa County. In developing them one company spent $145,000 on a short steam railroad, another went to great expense to build a gravity railroad, and all invested large sums in cutting some of the longest mining tunnels in the state and in buying pumps and steam engines. The capital for this came chiefly from San Francisco, but partly from the East.[19]

Coal mining, unlike the speculation in silver, copper, and petroleum, had the justification that, if successful, it would supply a material essential to the economic life of the state. The same could be said of California's most unique mineral industry: quicksilver. This was a branch of mining that California shared with very few other parts of the globe. For many centuries the Western world had derived its mercury from three great mines: the Almaden, in Spain, the Idria, in Transylvania, and the Huancavelica, in Peru. Then, in 1845, a Mexican officer discovered a large deposit of cinnabar ore in the hills south of San José, between the bays of San Francisco and Monterey. This he named New Almaden.

Lacking funds to develop the claim, the Mexican and

[19] *Ibid.*, March 11, 1865, January 18, 1868, March 6, 1869, April 27, 1872; Cronise, *Natural Wealth*, pp. 158-159; Browne, *Report* (1868), pp. 234-236; J. P. Munro-Fraser, *History of Contra Costa County, California, Including its Geography, Geology, Topography* (San Francisco, 1882), pp. 131-132.

his partners sold their rights to a company in which English capital was at first dominant, but in which control was soon shared with several wealthy San Franciscans. Preliminary attempts at development began as early as 1846 and 1847, and a small amount of mercury was secured in 1848. The mine did not come into production on a significant scale, however, until two and a half years after Marshall's discovery at Coloma. The Gold Rush gave the New Almaden company a huge local market for their product, and, conversely, after 1850 New Almaden was able to provide the gold miners with an indispensable aid to their operations.[20]

The New Almaden company had what was virtually a North American monopoly until 1858. In that year a virulent legal battle for ownership of the mine gave rise to a federal injunction which locked up the entire property until 1861.[21] Quite aside from the merits of the case, this judicial action injured directly every miner in California, and with the opening of the Comstock Lode and the other new areas, it affected the whole region west of the Great Plains. In 1850, when the New Almaden

[20] George F. Becker, *Geology of the Quicksilver Deposits of the Pacific Slope*, U. S. Geological Survey, *Monographs*, XIII (Washington, 1888), 1-50; George E. Dane, "Introduction" to Moerenhout, *Inside Story*, pp. v-ix; San Francisco *Weekly Alta California*, April 5, 1851; "The Quicksilver Mine of New Almaden," *Hutchings' Illustrated California Magazine*, I (1856-57), 98-99; Cronise, *Natural Wealth*, p. 590.
[21] Yale, one of the lawyers in the case, reviews it in his *Legal Titles*, pp. 333-336. A modern study is: Milton H. Shutes, "Abraham Lincoln and the New Almaden Mine," *California Historical Society Quarterly*, XV (1936), 3-20.

company was just beginning to put its product on the market, mercury had cost $99.45 per flask. Under the influence of the New Almaden's output, the rate had fallen gradually until it had reached the low level of $47.83 in 1859. During the year that followed the issuing of the restraining order, it climbed upward again to $63.13.[22]

This high price had one beneficial result: it led to a sudden burst of interest in prospecting for and developing additional quicksilver mines. During the years in which the New Almaden was closed, operations were begun at three mines in the area southeast of San Francisco Bay, not far from the New Almaden itself: the New Idria, Guadelupe, and Enriquita. At almost the same time several mercury mines were "located" north of San Francisco Bay, in a hilly region embraced by parts of Napa, Lake, and Sonoma counties. The most important of these was the Redington.

When the new mines began coming into production in the later sixties, the supply of quicksilver threatened to get ahead of the demand. Just as the closing of the New Almaden had stimulated an interest in quicksilver in California, so it had encouraged quicksilver firms the world over to increase their output. Now, with the New Almaden once more in the field, the producers had to think of adjusting the supply to the market. By that

[22] Walter W. Bradley, "Quicksilver Resources of California," California State Mining Bureau, *Bulletin,* no. 78 (May 1918), p. 11. But the retail price to the miner was higher.

time control over the New Almaden, the New Idria, and some of the smaller mines had fallen into the hands of a group of wealthy men who maintained intimate business relations with the Bank of California and with Barron & Co., of San Francisco, the chief distributing agency for mercury. With the might of the Bank of California hovering in the background, the owners of the Redington mine, which constituted the most dangerous "outside" threat, were admitted into the inner councils of Barron & Co. and the New Almaden and New Idria proprietors.

By that means a "combination" was formed. Thereafter—throughout the late sixties and until April 1873—the output of the California mines was arbitrarily regulated, and the price of mercury was maintained at profitable levels. Any surplus above far western consumption was exported to other parts of the world, often at rates below those preserved in California.[23]

At the time when this monopoly was established, New Almaden was already the principal producer of quicksilver in the Western Hemisphere and was rivaled in importance in the whole world only by old Almaden, in Spain. When the several California mines were brought

[23] The formation and operation of the combination can be pieced together from: *Mining and Scientific Press*, May 18, 1867, January 18, 1868, January 16, 1869, May 4, 1872; Rossiter W. Raymond, *Mineral Resources of the States and Territories West of the Rocky Mountains* (Washington, 1869), p. 10; Browne, *Report* (1868), p. 263; C. N. Schuette, "Quicksilver," U. S. Bureau of Mines, *Bulletin*, no. 335 (Washington, 1931), pp. 126-129; and the table of prices cited in the preceding footnote.

into a single orbit, they wielded, collectively, a power that reached across half the earth. The United States mining commissioner went so far as to say that "the quicksilver trade of the world is substantially an armed truce between Spain and California." [24]

When opportunity offered, the truce was broken, such as on the occasion when the California combination "dumped" their surplus in China at cut-rate prices, in order to wrest control of that market from the Rothschilds, who held the lease on old Almaden.[25]

The value of the total yield of New Almaden prior to the year in which the monopoly was formed was estimated at about $20,000,000. The mine was backed by the huge theoretical capital of $10,000,000. It had an annual expense account that averaged over $700,000 and a staff of employees that numbered about 1,200. The New Idria and Redington mines, with much smaller capacities, employed, respectively, 300 and 200 men.[26]

This profitable monopoly finally came to an end in April 1873. The agreement under which the combination was operating expired at that time, and the owners of the Redington mine refused to renew it. The New Almaden company had encountered a body of low-grade ore that had reduced their output by more than two-thirds, with the result that a shortage of mercury

[24] Raymond, *Mineral Resources* (1869), p. 10.
[25] *Ibid*. A table of quicksilver exports is given in John J. Powell, *The Golden State and its Resources* (San Francisco, 1874), p. 82.
[26] *Mining and Scientific Press*, January 18, 1868.

was threatening where once there had been a surplus. Prices had been rising for some time past. In the belief that prices would go higher if left alone than they would if regulated by monopoly control, the Redington proprietors decided to trust to fair competition. At the start of 1874 the United States commissioner was able to report that "the 'monopoly' so long existing in this product exists no longer; and the price is regulated by the well known laws of supply and demand." [27]

Not all of the time and money of adventurous men went into attempts to win fortunes with a special mineral like quicksilver or with minerals new to California like silver, copper, and petroleum. Gold still had its glamorous appeal, but in the traditional districts too many of the promising claims were too well-known to be available save by purchase. The later gold "excitements" and "rushes" were, of necessity, confined to the fringes of the gold region—to the remote or inaccessible areas that had been neglected hitherto.

Kern River, for example, experienced a revival that partook of most of the characteristics of a small boom. After its unfortunate false start in 1855, its name was in bad odor for ten years. Most men avoided it, and those who had the courage to try its potentialities were deterred by the rugged mountains that walled it off from

---

[27] Rossiter W. Raymond, *Statistics of Mines and Mining in the States and Territories West of the Rocky Mountains; being the Sixth Annual Report* (Washington, 1874), p. 29, and cf. pp. 27-28 on why the combination was dissolved.

the world. A change in the general attitude towards it came only in the middle sixties, when hard times clouded the Comstock Lode's attractiveness. Then finally a strong current of immigration turned towards Kern River once more.

After one fair-sized town and a number of small ones had sprung up, Kern was declared populous enough to be organized as a separate county. Some placer mining was done, but the thriving industry was the comparatively inexpensive business of "locating" quartz claims for speculative purposes. A few out of the many claims succeeded in winning financial support from San Francisco and the East. Most remained undeveloped. With engaging expansiveness the county asserted in 1868 that it had seventeen quartz mills and that twelve hundred persons were engaged in mining.[28]

Meadow Lake had a similar history, save that it ended its life more disastrously. Though a part of Nevada County, Meadow Lake was situated at so high an altitude—eight thousand feet—and in so rocky and snowbound a part of the Sierras that it was virtually unknown until the summer of 1865. Times were very dull just then over in Washoe, and when the whisper of a rumor arrived that quartz ledges had been found in the lofty Meadow Lake area, a crowd of restless miners started westward from Virginia City. They were soon joined

[28] *Mining and Scientific Press*, September 23, 1865, June 9, 16, 1866; Cronise, *Natural Wealth*, pp. 117-119.

by men from all parts of the state of Nevada and from the central mining section of California. It was said that three thousand people visited the place before the snow fell in late September, and that one hundred and fifty buildings were erected.[29]

Although no actual mining was done during 1865 and none of any significance in 1866, the rush continued through the greater part of the season of 1866. The rocky countryside was plastered with claim notices, and a group of speculators had the effrontery to organize a stock exchange to deal in the shares.[30] As winter approached, the rosy prosperity faded along with "the disappointed hopes of thousands of silly adventurers who failed to find mines of twenty dollar pieces already coined." [31] Then, and then only, did the expectant Vanderbilts pause to consider that the Meadow Lake gold-quartz was heavily impregnated with sulphurets, that the natural environment was a very difficult one, and that capital was not rushing in to Meadow Lake. The newly built "city" was deserted as rapidly as it had been populated, and a few years later not a living soul was to be seen on its once busy streets.[32]

---

[29] Its story has been told by the editor of its little newspaper: Frank Tilford, "Sketch of Meadow Lake Township," in *Bean's History*, pp. 305-312.

[30] *Meadow Lake Morning Sun*, June 6, 1866. This newspaper was advertised to be a daily! Its second issue was a weekly, and its demise soon followed.

[31] *Mining and Scientific Press*, October 6, 1866.

[32] *Ibid.*, February 17, 1872.

If this be taken as indicating that it was only the penniless who were harebrained, the history of Bodie should be sufficient to efface that impression. Bodie was a district on the eastern side of the Sierras, just north of Mono Lake and only a few miles from the Nevada boundary. Gold was discovered there in 1859 and a district organized in 1860. During the next two years the original locators sold out because of lack of capital. A consolidation of the claims was effected so that the incorporated Bodie Bluff Consolidated Mining Company could be formed, with Leland Stanford as president and with a capital stock of $1,110,000. When this proved a financial failure, it was succeeded by an even bigger concern, the Empire Company, of New York, which had a "paper" capital of $10,000,000 "and the modest privilege of increasing it to fifty millions." [33] Through gross extravagance this company also failed, and Bodie joined Meadow Lake in the ranks of the ghost towns. Not until a dozen years had passed did a sheer accident reveal ore chambers so rich that Bodie became, in the late seventies and the eighties, one of the most thriving mining towns in the Far West.

As a similar illustration of the gullibility of those who controlled capital, one could cite the famous diamond

[33] Wasson, *Bodie and Esmeralda*, p. 7. The story of Bodie is told in Wasson's pamphlet; also in an address by Grant H. Smith, "Bodie, the Last of the Old-Time Mining Camps," *California Historical Society Quarterly*, IV (1925), 64-80, and F. W. McIntosh, *Mono County, California, the Land of Promise for the Man of Industry* (Mono County, California, 1908).

hoax of 1872. Some unscrupulous California speculators purchased coarse uncut diamonds in Europe and "salted" them in a remote district of the Southwest, variously reported to be in Arizona, New Mexico, and Colorado. They then managed affairs so skillfully that a noted and entirely honest San Francisco mining expert was induced to make a hasty examination of the purported diamond field. He declared it genuine, but added some cautionary remarks about the inadequate nature of his examination. Disregarding the warning, some of the most prominent men in San Francisco, including the heads of the Bank of California and the London and San Francisco Bank, hurriedly formed a company to exploit the new treasure trove. Among their associates they had the Pacific Coast agent of the Rothschilds and so well known a public figure as General George B. McClellan. It was only after these optimists had been bilked of several hundred thousand dollars that an independent survey exposed the "diamond field" as an outright fraud.[34]

Looking back over this list of "excitements," one is struck by the very high percentage of failure and sheer waste. The quicksilver mines alone yielded as they were expected to, and even there it was not, in most cases, the original locators who benefited. The other manias cost

[34] The hoax can be traced in San Francisco *Daily Alta California*, August 1, 2, 4, 19, October 14, 16, 28, November 26, 1872; *Mining and Scientific Press*, August 3, 10, 24, September 7, November 2, 16, 30, 1872; Asbury Harpending, *The Great Diamond Hoax and Other Stirring Incidents in the Life of Asbury Harpending*, ed. by James H. Wilkins (San Francisco, c. 1915), pp. 195-264.

much and repaid little. What Senator Stewart said of gold and silver mining was true of them all: "Mining for the precious metals is prosecuted at a greater average loss than is suffered by any other business in the United States." [35]

Why, then, is it important to tell of these manias? Because this was the way that men spent their lives. Thousands of individuals committed their persons and their fortunes to these hopeless struggles for wealth. The lonely prospector and the restless miner were by no means the only victims. On the contrary, people of almost all types might be found at the short-lived boom towns. An observer gave this description of Kern River's population:

Stock-brokers, surveyors, lawyers, gamblers, dentists, prospectors, teamsters, French, Italians, and all kinds and characters of men, throng the streets from morning until night, and through the night until morning again, buying, selling, gambling and drinking—a perfect *fac simile* of Virginia City in 1861 and 1862. [36]

Behind those who actually participated in the rushes were the city folk who made their contribution in dollars. They, too, were of many types. There were financiers like Louis McLane and the president of the London and San Francisco Bank, mining magnates like Hearst,

[35] Committee on Mines and Mining, Report, *Senate Report*, 52 Cong., 2 sess., no. 1310 (February 18, 1893), p. 4.
[36] *Mining and Scientific Press*, September 23, 1865.

ministers and educators like the Reverend Henry Durant, and the "little fellows" like Henry George and Frank Leach. All had seen a few fortunes made in California in the fifties and in Nevada in the sixties, and they took their desperate chance in the hope of duplicating the success of others. A Senate committee has well stated:

In any other vocation in life so large a percentage of loss would destroy the industry. But the great prizes occasionally won inspire the ambition and incite the hopes of a race of men naturally adventurous; and so the work goes on and the business will continue with various successes and failures so long as the precious metals are used as money. But, in the opinion of your committee, mining for the precious metals has been and always will be prosecuted at great loss, sacrifice, and toil.[37]

[37] Committee on Mines and Mining, Report, *Senate Report*, 52 Cong., 2 sess., no. 1310, p. 5.

XVI

MATURITY
1860-1873

When compared with the hectic pursuit of sudden fortune in distant mines and new minerals, California gold mining of the later era seemed as stable as a grocery business. For the California gold region the days of wild excitement were past. The number of districts, the number of miners, and the number of types of mining had all been reduced greatly. The problem now was to secure funds and apply technical skill so as to make a small number of mines pay richly under the application of established processes.

Throughout the traditional gold area, the first half of the sixties was a dull period during which men's attention was focused on the greater California that was arising beyond the state's political boundaries. It was, also, a period during which drought and flood made operations difficult. A revival came finally in the late summer of 1864. It came not because better climatic conditions had returned, for indeed heavy rains did not fall until November of that year. Rather, the cause of improved conditions in California was the arrival of hard times in Nevada.

After five years, the flush days on the Comstock Lode

finally came to an end in the spring of 1864. The ruling prosperity had been based on the rich yield of superficial deposits. In 1864 these were approaching exhaustion. When the hundreds who did their mining through the stock market sensed trouble, disaster struck. In May quotations on Washoe stocks were declining. In June they were crashing abruptly before an uncontrollable public panic.[1]

Closed mines and unemployment became general in Nevada. Financial hardship became widely prevalent in San Francisco and other cities, since "the great majority of the shares were held in California."[2]

Washoe's loss was the California gold region's gain. A few months after the beginning of the stock market panic, the mining-town newspapers began reporting the return of many of their former citizens who had departed so optimistically for Nevada a few years earlier. Before the end of the year Nevada County was feeling the lift of better times. Elsewhere the home-coming wanderers were not able to do much towards restoring prosperity until after the rains had come in November and December.[3]

With the coming of 1865 the reports of returning citizens and reviving hope became more general, and before

---

[1] Lord, *Comstock Mining*, pp. 181-182; stock market reports in *Mining and Scientific Press*, April 16, May 14, 21, 28, June 4, 11, 18, July 2, 1864.
[2] Lord, *Comstock Mining*, p. 181.
[3] *Mining and Scientific Press*, August 13, 27, October 22, November 5, 1864, January 7, 1865; Grass Valley *Daily National*, August 23, 1864.

many months had passed the old California districts found themselves basking in the gentle warmth of an Indian summer. To a limited extent this return from abroad continued for several years. It was said that miners were coming back from places as distant as Montana, Idaho, and British Columbia.[4]

This renewed interest in California came precisely at the moment in which the state's yearly output reached its permanent annual average, and that fact implied that mining had completed its transition from a multitude of types to two: hydraulic and quartz. The older forms, such as working the gulches, bars, and river beds, had been relegated to the Chinese and to a stubborn band of white miners who preferred small operations of their own to participation in great projects owned by others. Tunneling was still carried on with great energy at places where conditions were unusually favorable, such as at Table Mountain, in Tuolumne County, at Forest Hill, in Placer County, and at several points in Calaveras and Sierra counties.

All of these types were now overshadowed by hydraulic and quartz mining. As between the two, the former was by far the more important, if measured in terms of results. A statistician has claimed that from 1861 through 1870, 90 per cent of the state's gold was derived

---

[4] *Mining and Scientific Press*, January 14, 21, February 11, 25, April 1, July 8, September 30, October 28, November 25, December 9, 1865, March 9, 1867.

from placers, and from 1871 through 1880, 70 per cent.[5]

Large-scale hydraulicking was of necessity confined to those counties of the central section that lay north of El Dorado, since the best of the deep gravels and water-ditch facilities were located there. Quartz mining was attempted in all of the Mother Lode counties and throughout the central section, but notable success was achieved only in the vicinity of the two centers at which lode mining had won its chief victories in the fifties: Grass Valley, Nevada County, and the towns in Amador County's ten-mile piece of the Mother Lode belt.

Outside of these two areas there were groups of adjacent quartz mines that experienced short periods of prosperous activity, and there were individual mines that could show long records of profitable operation. Nowhere else, however, were there clusters of adjacent mines that remained in production with sufficient regularity to support good-sized communities. Apparently the greater part of the lode mining region was characterized by ore bodies that either varied from low-grade to marginal in value, or, when high-grade, were too small in size to justify continuous operation.[6]

The area that might well have been an exception to

[5] Hill, "Historical Summary," U. S. Bureau of Mines, *Economic Paper*, no. 3, p. 5.

[6] Cf. Knopf, "Mother Lode System," U.S.G.S., *Professional Paper*, no. 157, pp. 7, 26-31; Carl E. Julihn and Frederick W. Horton, "Mineral Industries Survey of the United States: California. Calaveras County. Mother Lode District (South)," U. S. Bureau of Mines, *Bulletin*, no. 413 (Washington, 1938), pp. 94-97.

this generalization was that comprised within the 44,000-acre Mariposa Estate. This was, of course, private property, owned at first by John C. Frémont, under a Spanish grant, and later by two successive incorporated companies. The first large-scale quartz operations in California were conducted on the estate, and for many years its mines, mills, workshops, warehouses, wagon roads, railroads, and water system were collectively the greatest single mining investment in the state save for the New Almaden quicksilver mine.[7]

Nevertheless, the estate was perennially in debt, and on several occasions mining within its limits had to be suspended entirely. The cause of the deficiency was partly that the ores of its three leading mines "decreased abruptly in value after a moderate depth had been attained," and that sulphurets and a shortage of water raised problems. On the other hand, it is quite apparent that work of all kinds was hampered by absentee ownership, speculative management, notorious inefficiency in milling the ores, and a notable absence of intelligent long-term planning.[8]

The Mariposa Grant was thus not as much of an ex-

[7] Allan Nevins, *Frémont, Pathmarker of the West* (New York, 1939), pp. 371, 395-396, 461-465, 583-587; San Francisco *Daily Alta California*, December 29, 1858, February 5, 1872; *Sacramento Weekly Union*, January 1, 1859; *Mining and Scientific Press*, January 21, 1865.

[8] The quotation is from Knopf, "Mother Lode System," p. 83; on sulphurets, see Julihn and Horton, "Mineral Industries Survey. . . . Tuolumne and Mariposa," p. 96; on other factors, see Browne, *Report* (1868), pp. 22-23; *Mining and Scientific Press*, March 25, 1865, January 14, 1871; San Francisco *Daily Alta California*, April 11, 1870.

ception as one might have anticipated, even though it was claimed in 1872 that twenty million dollars' worth of gold had been extracted from it.[9] Certainly it was never, after the early days, as important to the technical development of the quartz industry as either Grass Valley or Amador County.

In the latter two areas, and in some of the other good lode districts also, the slow forward movement of the industry was greatly accelerated by the return from Washoe of many seasoned men who had once served an apprenticeship in the California quartz mines, and now were coming back to ply their trade in the school in which they had learned their first lessons.[10]

It was fortunate that these "Washoeites" decided to return to the western side of the Sierras. Comstock mining had gone beyond the farthest limits of California progress in regard to operations underground. Comstock shafts and tunnels were longer and equipment was better.[11]

Once back in California, the former Comstockers quickly demonstrated the necessity for sinking shafts to greater depths than Californians had believed financially feasible, and when they proved their point on that score, they automatically forced Californians to give attention to methods of underground ventilating, to adopting bet-

[9] San Francisco *Daily Alta California*, February 5, 1872.
[10] *Grass Valley Daily National*, October 14, 18, 1864, January 16, 1865; *Mining and Scientific Press*, November 5, 1864.
[11] Lord, *Comstock Mining*, pp. 220-227.

ter hoisting equipment, and to replacing hemp rope with steel or iron cables.[12]

The decision to go deeper into the earth also made it necessary to reconsider the type of explosive used in breaking up the heavy bodies of underground rock. Until the middle sixties gunpowder had been used almost entirely. It had been secured chiefly from the agents of the Du Pont and Hazard companies, who controlled the far western market and sometimes exacted high prices. During the Civil War years the high prices and the uncertainty of obtaining gunpowder from the war-torn East led to the establishment of two California powder works.[13]

By bad luck, the local manufacturers had hardly put their plants into operation before attention began to shift from the old-fashioned gunpowder to the new and deadly nitroglycerin that had just been developed by Nobel, in Sweden. This proved too dangerous a tool in its original liquid form, but when transformed into a solid, it passed into widespread use under the name of dynamite, or "Giant Powder." [14]

With better explosives and better underground equipment, it was found that mines which had ceased to pay when their shafts had reached a depth of 200 or 250

[12] *Mining and Scientific Press*, February 13, 20, 27, 1864, April 21, December 29, 1866, January 12, 1867.

[13] *Ibid.*, January 16, April 16, October 15, 1864.

[14] *Ibid.*, December 23, 30, 1865, April 28, June 9, 1866, April 6, June 22, 1867, February 22, 1868, February 6, 1869.

feet could now be made to pay satisfactorily at twice that depth. During the sixties and seventies two of the best mines in the state, the North Star and Empire, both at Grass Valley, were extended down to about 1,000 feet on the incline. The Hayward mine, in Amador County, broke all records in California and perhaps all in the United States when it attained 1,350 feet in 1870.[15]

The improvement in underground techniques was paralleled by a slow bettering of surface methods. A few mines began to substitute mechanical rock breakers for the manual labor formerly employed to reduce the excavated auriferous rock to a proper size for the stamp mills. Others adapted the idea of the "Washoe Pan" to the needs of gold mining, in order to extract a larger proportion of the precious metal from the rock after it came from the stamp mill. A few began to give the chlorination process the attention it deserved.[16]

The first attempts to use the latter process had proved a failure, at the close of the fifties. The men who were experimenting with it became discouraged and went off to Washoe. "Here they gained new ideas in relation to the working of metals, and in 1860 they returned to re-

---

[15] *Ibid.*, January 12, 1867; William Hague and W. D. Pagan, "The North Star Mine, Grass Valley," *ibid.*, October 10, 1914; George W. Starr, "The Empire Mines, Past and Present," *ibid.*, August 4, 1900; on Hayward, see Raymond, *Statistics* (1870), p. 32. "On the incline" means following the dip of the vein.

[16] *Mining and Scientific Press*, January 7, 1865, August 10, 1867; Phillips, *Mining and Metallurgy*, p. 198.

sume their experiments, and the first attempt was a success." [17] The potentialities of the process were considerable, for contemporaries estimated that improved California quartz mills could save 75 per cent of the gold without chlorination, but 90 per cent with it.[18]

Even with so favorable a ratio, the process came into use very slowly. More than a half-dozen years passed before it was employed to any significant extent outside the locality of its California origin: Grass Valley and Nevada City. Amador County seems to have been the second convert to the process. Elsewhere its adoption was a slow business. Certain machines complementary to the process were still lacking, and even with them there would have been doubt as to whether chlorination could be used profitably in any area save one in which the local mines had large quantities of sulphureted ores to be treated.[19]

Perhaps also, in this as in dealing with several other problems, the quartz industry was being guilty of complacency. Having achieved success in the fifties only after rejecting intriguing innovations, the quartz operators seem to have been reluctant to spend money on experimental work during the sixties, when their mines

[17] *Bean's History*, p. 127.

[18] *Mining and Scientific Press*, October 12, 1867; confirmed by *Bean's History*, p. 127, and Mason, *Amador County*, pp. 148-149.

[19] Cronise, *Natural Wealth*, p. 558; Phillips, *Mining and Metallurgy*, p. 196; Jackson *Amador Weekly Ledger*, January 2, 1869; *Bean's History*, p. 247.

were making available the funds required for trying out new methods.[20]

No such charge of conservatism could be made against hydraulic mining. That branch of mining strode ahead during the sixties and early seventies just as it had during the fifties. The worst problem to be solved was that of increasing the water pressure without bursting the short length of "crinoline" hose that served the miner as his means of changing the direction of the stream of water.

In the early sixties the importance of the question became great, because by that time the hydraulickers had well-nigh exhausted the low, soft banks of gravel. They were now faced by the necessity of attacking high hills of resistant material. To do so they must use a stream of water of sufficient power not only to disintegrate the hard gravel, but also to disintegrate it at a considerable distance. The higher the hill of gravel, the further back the hose-man had to stand if he were to avoid being caught under the debris when his jet of water undermined it.[21]

The need was met by a series of complementary inventions that collectively made possible a unit that was fashioned entirely of metal and yet had as much flexibility as the canvas hose. One inventor suggested attaching the

[20] Cf. *Mining and Scientific Press*, October 20, 1866; William H. Storms, "The Mother Lode Region of California," California State Mining Bureau, *Bulletin*, no. 18 (October 1900), pp. 26 ff.
[21] *Mining and Scientific Press*, September 21, 1863.

nozzle to a swivel, in order to permit horizontal changes in direction. Others devised globular and "knuckle" joints, and adjustable gears, by which the vertical aim could be controlled. Simultaneously a saving in labor was achieved by mounting the improved discharge piece upon a firmly fixed tripod, so that a single person could operate the unit, instead of requiring two strong men to brace themselves and cling to the nozzle in fireman's style.[22]

A second problem was posed by the fact that the masses of auriferous gravel often occurred above a basin-shaped depression in the bed rock. Since the richest gravel always lay at the bottom, it was essential to wash out the entire mass, and that could be done only by cutting a drainage tunnel that would tap the lowest part of the rock basin.[23]

The slowness and expense of cutting tunnels through solid rock inevitably turned men's thoughts to the possibilities of contriving a power-driven drill capable of boring through any material. Californians read with interest of the drilling machines that were being developed in Europe for use in the Mount Cenis tunnel, and in the East for the Hoosac tunnel. One California miner had the originality to design and patent a machine of his

[22] *Ibid.*, December 7, 1867, May 2, 1868, July 9, September 10, October 1, 1870; Bowie, *Practical Treatise*, p. 50; Rossiter W. Raymond, *Statistics of Mines and Mining in the States and Territories West of the Rocky Mountains* (Washington, 1872), pp. 62-65.

[23] *Mining and Scientific Press*, October 19, 1867, January 11, 1868.

own. But no real progress was made until a group of Californians and easterners joined in adapting to the needs of western mining the diamond-studded drill that had been invented in 1863 by the Frenchman, Leschot. It was found that this instrument was not only an invaluable aid to all tunneling and blasting operations, but that it could also be used to "prospect" a claim, since it would cut through solid rock and bring out specimens in the form of thin cylinders of auriferous material.[24]

While the diamond drill and the improved hydraulic unit overcame the worst difficulties of the hydraulic industry itself, there remained the problem of providing for the related interests of the water companies. That was a field in which the conditions of the later era were anything but favorable to prosperity. In the fifties there had been a universal and insatiable cry for water; in the sixties the hydraulic districts were the only ones in which the demand for copious quantities still continued.

Over the greater part of the mineral region the ditch companies found their consumer market disappearing and their debts mounting. Their ditches had been constructed in a day of high wages and high costs, and that fact meant paying interest on a larger investment than was justified by later conditions. The technical work had been done by engineers whose ideas were often faulty and unduly costly. Since there had been a maxi-

[24] *Ibid.*, February 4, 1865, March 27, October 9, 1869, January 8, November 5, 1870, June 3, 10, 1871; Blake, *Notices of Mining Machinery*, p. 1.

mum use of wooden flumes, the annual expense for repairs was heavy. Later experience showed that a large yearly saving could be made by using iron pipe.[25]

A few companies tried to bolster their revenue by hiring Chinese to work over the auriferous ground served by the company. One concern tried to build up a new manufacturing center in the Sierra's foothills, in order to create a market for water. Others sold water to mountain farmers for irrigation.[26]

All of these measures were mere palliatives. The ditches and flumes had been built to serve miners who were willing to pay high duties. They could not be operated profitably on any other basis. By the later sixties reports of abandoned lines were becoming frequent, and save for the deep gravel areas, few companies were able to sell all the water they could supply. Where once they had been the most favored form of investment, the water systems came now to be regarded as poor property. The federal mining commissioner reported that "not less than $20,000,000 have been invested in the mining ditches of California," but "their present cash value is not more than $2,000,000." [27]

In both quartz and hydraulic mining the dominant trend throughout the sixties and seventies was adverse

[25] Browne, *Report* (1868), pp. 180-182; Hittell, *Resources of California*, pp. 303-306; *Scientific Press*, January 7, 1871.

[26] Browne, *Report* (1868), pp. 181, 196-199; *Scientific Press*, February 5, 1870.

[27] Browne, *Report* (1868), p. 180.

to the continuance of independent, small-capital enterprises. The quartz miner found that "with the increase of depth, the difficulties of mining increase and the quality of ore often changes, so that more extensive or complicated machinery is required and more capital is necessary."[28]

The hydraulic miner encountered similar expenses when he expanded the size of his operations, and he discovered that it was not financially justifiable to cut a bed rock tunnel, or to bring in large quantities of water, unless he could count upon washing hundreds of tons of auriferous debris, so as to spread his initial investment over many units of material handled. For that reason it became the normal preliminary to large-scale work to consolidate the many small, individual claims into a single piece of property.[29]

Similarly, as a necessary consequence to the increasingly ambitious proportions of hydraulic operations, the water companies had to be prepared to supply water in large amounts to those districts in which hydraulicking was the prevailing method of mining. That meant building big dams and reservoirs, and constructing pipe lines of large capacity, both of which steps required an extensive capital outlay.[30]

[28] *Scientific Press*, October 15, 1870.
[29] *Mining and Scientific Press*, October 19, 1867, January 11, 1868, January 2, 23, March 13, 1869; *Scientific Press*, July 9, September 24, October 1, November 12, 19, December 3, 1870.
[30] Cf. Rolfe, "Mines and Mining," *Bean's History*, pp. 65-72; *Scientific Press*, September 3, October 22, 1870, July 8, 1871.

In every field, then, the day of small-capital mining was past, and the time was ripe for the advent of "heavy" finance. When Nevada reached this same point in her history, she unhesitatingly turned to the incorporated company and the stock market. California showed greater discrimination.

In March 1863, the *Mining and Scientific Press* compiled a list of incorporated mining companies that had their principal business offices in San Francisco. The *Press* was able to enumerate 203 companies. Of these, 97 were engaged in mining on the Comstock Lode and in the Humboldt district of Nevada, and 62 were conducting their operations near Esmeralda, on the Nevada-California boundary, or in the arid region south of Esmeralda; 14 were mining in the northern provinces of Mexico. Only 30 companies were described as operating "California mines," and of these 19 were in a single district in the new mineral area in the eastern part of the state. That left a maximum of 11 companies that may possibly have been interested in mines in the traditional California districts.[31]

This was an accurate indication of the relation of corporate capital to the older California areas. In 1865 the Grass Valley *National* declared that "we have but three incorporated mining companies in this Township," and it bitterly opposed proposals to establish a stock exchange at Grass Valley, on the grounds that such a

[31] *Mining and Scientific Press*, March 30, 1863.

step would only lead to speculation and fraud.[32] This conservatism on the part of an old town like Grass Valley stood in sharp contrast to the practice in the new communities along the eastern edge of the state. Long before substantial progress had been achieved in the silver districts of Alpine County, an "Alpine Stock and Exchange Board" was organized, and before any significant development work had been done at Meadow Lake, that raw young camp had its "Excelsior Stock Board."[33]

This does not mean that "outside" capital was not finding its way into the older mines, but rather that the movement of funds into Nevada, Amador, and the other successful counties was of a slower and more conservative nature. In October 1864, the Grass Valley *National* declared that "San Francisco and Eastern capital are liberally invested in the mines of Grass Valley," and a year later the *Mining and Scientific Press* said of California quartz mines generally: "A very large proportion of the late mining sales, it will be noticed, have been made to San Francisco capitalists, who, as a general thing, are buying for permanent investment."[34]

Quartz mines seem to have been sold to San Francisco buyers more frequently than hydraulic claims, but several of the largest hydraulic projects were financed

---

[32] Grass Valley *Daily National,* January 27, 1865, also January 24.

[33] *Grass Valley National,* March 19, 1864; *Meadow Lake Morning Sun,* June 6, 1866.

[34] Grass Valley *Daily National,* October 14, 1864; *Mining and Scientific Press,* October 14, 1865.

primarily by San Francisco interests. In making purchases of mining property California investors tended to operate in groups, since no one man cared to risk the large amount required for a good mine. Often members of the group acted in their personal capacity. Occasionally they used the device of an incorporated company in which the stock was never sold to anyone outside the group.

Not until 1867 was any California mining stock quoted in the daily San Francisco market reports, and not until the latter part of that year was an important California stock placed in the market. At that time the rich Hayward Amador mine was incorporated with a capital of $1,480,000, and its stock was offered on the Exchange Board. Shares in other incorporated California mines, offered for general sale, appeared so slowly that the leading mining journal waited until December 1868 before it provided a listing of California stocks separate from those of Washoe, and even then it gave quotations for only four.[35]

At as late a date as January 1874, the *Mining and Scientific Press* declared:

It is a significant fact to mention that the majority of the good quartz mines of the State are in private hands and pay well enough in themselves, without the necessity of the owners having recourse to stock jobbing operations. Several of the California street operators own quartz mines and mills in this State

[35] The Hayward mine was reported in the *Mining and Scientific Press*, October 26, 1867; California stocks listed in *ibid.*, December 5, 1868.

which pay them steady profits from one year's end to the other, and they are shrewd enough to keep such properties in their own hands. Even when they are incorporated the stock is held by a few owners who have no desire to sell.[36]

Long before that date, city capital became an important factor in the ownership of California mining properties of all types. There were still many miners, however, who felt that urban financiers were unreasonably reluctant to trust their dollars in the mines of their own state, and to them the criticism seemed justified by the eagerness with which English interests were invading the California investment field.

In the middle fifties, after suffering heavy losses in the early quartz mines, "The English turned their backs upon California as a field for profitable investment." [37] Now, with California mining on a firm basis, with a federal law granting private title, and with charges for labor, equipment, and transportation all greatly reduced, the English began to reënter the California arena. Practical miners often preferred to deal with them, because the English were content with a smaller rate of interest, and because the miners "regarded mining capitalists here [i.e., in San Francisco] as merely speculators, and were rather afraid of many of them." [38] Just how much capital flowed into the Sierra foothills from the British Isles it would be impossible to say. Reports of British purchases

[36] January 17, 1874.
[37] Ibid., December 2, 1871.
[38] Ibid., January 1, 1870.

and negotiations were constantly appearing in the press, but there is no way by which one can measure the actual sums involved.

The presence of eastern capital in California mining investments was not mentioned so frequently. There were some great projects, such as a huge Nevada County water company that was bought by New York interests in 1865, and was incorporated for two and one quarter million dollars.[39] Others were referred to from time to time, but one gains the impression that the bulk of the eastern funds was going into Colorado and Nevada rather than into California.[40]

The same factors that made it necessary to bring "heavy" finance into mining also created a need for a more scientific direction of operations. As the quartz miners went deeper into the earth, and as the hydraulickers attacked larger and yet larger hills, they found themselves facing more complicated geological problems and using increasingly intricate machinery. At the same time, before a "capitalist" made an important investment in a mine, he wanted to have a full report on the property by a competent engineer or geologist.

The realization that changed conditions were coming into existence forced contemporaries to acknowledge

[39] Rolfe, "Mines and Mining," pp. 65-68.
[40] Cf. Hollister, *Mines of Colorado*, pp. 131-133, and Samuel Bowles, *Across the Continent: A Summer's Journey to the Rocky Mountains, the Mormons, and the Pacific States, with Speaker Colfax* (Springfield, Massachusetts, 1865), pp. 145, 154, 157.

that rule-of-thumb methods were no longer sufficient. Traditionally the working miner had been scornful of science. His disdain was well expressed by the Sacramento *Placer Times* when it said in 1849:

The mines of California have baffled all science, and rendered the application of philosophy entirely nugatory. Bone and sinew philosophy, with a sprinkling of good luck, can alone render success certain. We have met with many geologists and practical scientific men in the mines, and have invariably seen them beaten by unskilled men, soldiers and sailors, and the like.[41]

Even at that early period there were some who had a more just appreciation of the value of learning. In 1850 and 1851 the *Alta California* began urging the inauguration of a state geological survey, and in January 1852 the governor of California endorsed the proposal. On two different occasions legislative committees recommended that either the state or the federal government take action. When, during the latter part of 1851, an enthusiastic amateur, Dr. John B. Trask, fitted out a so-called geological survey, he was rewarded by the legislature with the title of state geologist and with a small appropriation. Trask did as much as could be expected of a person who had had little formal training and possessed only limited resources, but something more substantial was needed, and at the close of the fifties a

[41] San Francisco *Alta California* (steamer ed.), August 2, 1849, clipping Sacramento *Placer Times*.

widespread sentiment was developing for a genuine state-supported geological survey.[42]

A dozen newspapers, located in many different parts of California, went on record as favoring the idea.[43] The eventual inspiration for such an undertaking doubtless came from the many similar enterprises that had been in progress for two decades in the older parts of the United States. "Between 1833 and 1836 nearly every one of the Eastern and Middle, and several of the Western, States commenced their geological surveys."[44]

The more immediate motive for seeking a survey of California was the desire to advance the state's economic life, and especially to aid its mining industry. The North San Juan *Hydraulic Press*, a mining-town newspaper, summed this up concisely: "It is generally held that a thorough geological report, from a competent person, would place the immense resources of California in a more commanding position before the world's eye, besides resulting in a more systematic method of developing them."[45]

When some public-spirited men placed themselves at the head of the movement, the legislature of 1860 was

[42] On the *Alta*, see San Francisco *Weekly Alta California*, October 5, 1850, February 22, 1851; on the Governor, see California Senate *Journal*, 3 sess., 1852, p. 20; for committee reports, see Assembly *Journal*, 2 sess., 1851, pp. 1689-1702, and Senate *Journal*, 3 sess., 1852, pp. 659-665; Trask's reports were published annually by the legislature, 1853-1856.

[43] See list in San Francisco *Daily Alta California*, December 7, 1858.

[44] Josiah D. Whitney, *Geographical and Geological Surveys* (Cambridge, Massachusetts, 1875), p. 81.

[45] North San Juan *Hydraulic Press*, January 15, 1859.

induced not only to establish a state geological survey, but also to appoint as state geologist the best qualified man in America: Josiah D. Whitney, a product of Yale and of long study and field work in both Europe and America.[46] In view of the standards of most California legislation, this was properly regarded as a modern miracle, and there were those who feared it was too good to last. Their apprehensions were amply justified.

As Whitney himself confessed, it was the popular belief that the geological part of the survey would consist of the very practical work of prospecting for new mineral deposits. Nothing was further from Whitney's intentions. A thorough-going scientist, he proposed to make the survey a complete analysis of the state's natural history and physical characteristics. When he reduced the survey's findings to print, the volumes were devoted to palaeontology, botany, and conchology rather than to new deposits of gold.[47]

Moreover, when the advice of the survey was sought concerning specific mineral discoveries, Whitney and his subordinates won only enmity for themselves, because they were always honest and hence skeptical about supposed "sure things." At the time of the ill-fated petroleum boom, for example, the survey's cold-blooded disbelief brought down upon it the wrath of politically

[46] Edwin T. Brewster, *Life and Letters of Josiah Dwight Whitney* (Boston, 1909).

[47] Josiah D. Whitney, "State Geological Survey," *Overland Monthly*, 1st series, VIII (1872), 80-87.

powerful promoters who stood to lose if the boom collapsed.[48]

In practice, the state geological survey proved to be much less of a help to the California mining industry than had been anticipated. As a scientific project, it was of world-wide significance, but as an aid to a branch of economic life that badly needed scientific guidance, it did not come up to contemporary expectations. Eventually it was to be of assistance to the whole American mineral industry, for it served as the proving ground for both the men and the methods that went into the new United States Geological Survey when that organization was founded in 1879.[49]

Thanks to the devotion and unselfishness of its personnel, the California survey managed to eke out a spasmodic existence from 1861 until 1874. It then disappeared under the combined weight of politics, personalities, and public apathy.

The failure of the survey to prove of immediate benefit to mining did not mean the closing off of all avenues to scientific knowledge. While the idea of a geological survey was still nebulous, mining men were beginning a long struggle to establish a medium for collecting and disseminating technical information. For the moment they succeeded only in setting up a number of ephemeral

---

[48] Brewster, *Whitney*, pp. 245, 266-267, 293-296.
[49] *Ibid.*, pp. 291-312; and see *ibid.*, pp. 282-303, on the reasons for the Survey's demise, but see also *Mining and Scientific Press*, April 4, 1868, November 13, 1869; *Sacramento Daily Union*, March 19, 1872.

societies and bureaus that made tentative essays at encouraging discussion groups, compiling reliable data, and assembling reference libraries.[50]

Before the last of these local efforts died, the California and Nevada delegations in Congress undertook to secure the establishment of a national mining bureau that would do for the miner the same service that the new "Department" of agriculture was intended to perform for the farmer. Their endeavors were crowned with success in 1866 when an act was passed creating the office of United States commissioner of mining statistics. The act and the accompanying instructions from the secretary of the treasury specified that the commissioner was to assemble information in regard to every aspect of American mining, and was to publish his findings each year in a report. The first holder of the position was a Californian, J. Ross Browne, and all of the stout volumes issued during the ten year existence of the office were concerned primarily with mining west of the Rockies.[51]

It was realized that it was not sufficient merely to make technical data available. There was a need also for men with enough training to use effectively whatever

[50] *Grass Valley Telegraph*, November 27, 1855; "County Miners' Association," and "Another Quartz Mining Association," *California Mining Journal*, II (1857-58), 29; *Mining and Scientific Press*, August 24, 1861, November 11, 1865, February 17, 1866.

[51] The establishment of the office is discussed in *Mining and Scientific Press*, January 28, February 18, 1865, September 8, 1866, February 23, 1867; the secretary's instructions are in Browne, *Report* (1867), pp. 4-5.

new ideas might be suggested, and there was an even greater need for men capable of doing original experimental and fact-finding work. For four years the Far West's only mining trade journal campaigned editorially for the establishment of a school of mines in the Pacific states. A limited degree of success rewarded its efforts in 1866, when two of the struggling little California colleges agreed to offer courses and laboratory instruction in chemistry and metallurgy.[52]

That was hardly enough for the task at hand. Sensing that the Far West was not ready to finance an adequate teaching program, Commissioner Browne and Senator Stewart of Nevada sought for a time to induce Congress to establish a national school of mines, similar to the French École des Mines or to the famous royal academy at Freiberg, Saxony.[53]

Perhaps it was just as well that the Senate reacted unfavorably to this proposal. In 1862 Congress had passed the Morrill Act, the purpose of which was to endow institutions that would provide an agricultural and industrial education. Until December 1865, the California legislature made no attempt to avail itself of the offer

[52] See frequent editorials in *Mining and Scientific Press*, January 11, 1862-January 27, 1866, and on the college courses see *ibid.*, February 17, March 3, 1866.

[53] See *ibid.*, September 28, 1867, January 4, 25, 1868, and see two pamphlets, apparently compiled by Browne, *Considerations in Reference to the Establishment of a National School of Mines*, and *Opinions of the Press and of Eminent Public Men on . . . a National School of Mines* (Washington, 1867 and 1868, respectively).

implied in the act. Finally the legislature was stirred into action by the prodding of the agriculturists, who suggested the founding of an "Agricultural and Mechanical Arts College." Subsequently this was broadened so as to read "Agricultural, Mining, and Mechanical Arts College," and in this backdoor fashion mining finally entered the California educational world.[54] It is significant of the shift in the relative importance of the two that the once omnipotent mineral industry had to rely upon the strength of the farm vote to gain its desires.

The new institution, under the name of the University of California, made provision for professors of mining, metallurgy, and geology. It did not, however, open its doors until September 1869. For that reason it did not have time to produce a significant number of engineers until the seventies were well along in their course. In the meantime, the need for trained men was supplied by an increasing number of graduates of the European and the new eastern American schools.

The final establishment of a California training school was as symbolic as the transfer of mines to San Francisco and English ownership. The one meant that henceforth distant city dwellers would give the orders and take the profits; the other that in the future the medium through which decisions would be made would be the advice of men educated in the laboratory and the classroom.

[54] William W. Ferrier, *Origin and Development of the University of California* (Berkeley, California, 1930), pp. 42-46, 62-77, 256-275, 279, 314.

Mining had started as an adventure in which men engaged with little capital and less knowledge. Gradually it had become a business conducted upon the basis of common sense rules that had been learned through practical experimentation. Now it was being transformed into a modern industry in which the dominant figure was not the "honest miner" who dwelt in the Sierras' foothills, but rather the financier in San Francisco or London, and the highly paid consultant or superintendent who made of mining a science and a profession.

With the attainment of that stage, the industrial revolution was complete. Then truly could it be said that mining had come into its full development—that it had passed from the disorganized splendor of its youth to the ordered stability of its maturity.

XVII

FROM SUNDAY CARNIVAL TO LABOR WAR
1857-1873

It is perhaps inevitable that in a mining region the achievement of social stability should be postponed until after the days of great richness have passed. As long as flush times continue, mineral districts are bound to be a mecca for all the adventurers of the earth. Then, too, as long as extensive gold or silver deposits can be worked with comparative ease, there is little incentive to anyone to settle down and associate himself permanently with any one area, while there is much to encourage the restless to wander from one district to another, making money quickly in one place and spending it as speedily in the next.

In California the newspapers began to talk hopefully of better conditions at as early a date as 1852. It was alleged in that year that there were already signs of an increased number of women and children and a greater amount of benevolence, both of which trends contrasted favorably with the previous tendency to excessive masculinity and too frequent dissipation.[1]

Three years later a more definite indication was given that quieter conditions were approaching. In response

---

[1] *Sacramento Weekly Union*, January 10, 1852.

to a public demand, the legislature passed a law to suppress gambling, and another "to prohibit Barbarous and Noisy Amusements on the Christian Sabbath." Simultaneously the much derided temperance advocates, who had been tenaciously at work in California since 1850, succeeded in forcing upon the state an unsuccessful referendum on a prohibition law.[2]

Along with the quieting down of society went an improvement in the physical conditions amidst which men lived. A better diet was now possible, thanks to the development of local agriculture. Transportation facilities were more widely available. In the towns a few of the buildings began to be more substantial.

By a piece of tragic irony, this change for the better frequently came precisely at the moment that a particular town was reaching its greatest prosperity, and only a short time before it began the rapid tumble into decadence. Columbia, for example, was said in 1856 to be fast filling up with families. It was said that the lawless Mexican population, the bloody Sunday bear-and-bull fights, and the crowded gambling houses were all disappearing. Two years later a feminine visitor reported that Columbia had "agricultural improvements, which are now quite extensive and permanent in their character . . . beautiful homesteads, cosily nestled in flowering

---

[2] *Statutes of California*, 6 sess., 1855, pp. 50-51, 124-125; on temperance, see *Sacramento Illustrated* (Sacramento, California, 1855), p. 13, and *Sacramento Daily Union*, September 3, October 3, November 5, 1855.

gardens, and in the midst of growing fruit." [3] Even as she wrote, Columbia's best days were slipping into the past, and before the advent of the sixties Columbia's miners were looking to other districts for richer fields.

The achievement of social normalcy was, therefore, sometimes too late to benefit the towns and camps that needed it the most. But a more ordered society did make its appearance in the foothill towns in the later fifties. The change was strikingly described by a veteran Californian who revisited Sonora on a Sunday in October 1857 and climbed the hill above the town, while the church bells below pealed out their summons to worship.

Seven years ago I was here, and on the Sabbath day. That was such a Sabbath as was never witnessed out of California, or by any man except a pioneer Californian.—Then, instead of the music of the church bell, was heard the clangor of brass bands drawing victims to the gambling hells and liquor booths; instead of orderly men, virtuous women and innocent children, wending their way to the Sunday school and the sanctuary, the place was thronged with the devotees of dissipation, and the very hills echoed the discordant noises that proclaimed the absence of the civilizing presence of woman. Sonora was then a temporary mining camp, flooded with gold, gamblers and rum; and Sunday was the day of high carnival.

How different now! The church has taken the place of the gambling house—the family residence has superceded the grog shop,—gardens and shrubbery are where the rude brush corral once stood; women and children walk the streets once monop-

---

[3] "Our Visit to Columbia," *Hesperian,* I (1858-59), 88; the description of 1856 is from Heckendorn and Wilson, *Miners & Business Men's Directory,* p. 8.

olized by rowdyism of every grade. Those wild scenes are gone, and a Sunday in Sonora is marked by all the evidences of a high state of refinement and virtue, that characterize cities and towns of other countries, that have had a century to grow in.[4]

As further specific evidence of the change, one could cite the appearance in the mining towns of many institutions that were familiar throughout the eastern portion of the United States. For example, literary societies, library associations, debating societies, lyceums, and glee clubs began to spring up in an increasing number of communities. The thriving hydraulic town of North San Juan boasted a dancing school and, *mirabile dictu*, a soda fountain! The Odd Fellows, Masons, and other fraternal orders reported a great increase in membership.

Sometimes Californians themselves were not fully aware of the extent of the change going on around them. Henry S. Brooks, a keen-eyed editor, remarked in 1861: "I dare assert that . . . there is not one in a hundred who realizes the extraordinary revolution that the last few years has created. There are numbers who have worked on, half dormant to all but the accumulation of gold. . . ."[5]

Behind Brooks's assertion lay the vivid impression that he himself had received a short time before when he

[4] Stockton *Weekly San Joaquin Republican*, October 17, 1857, correspondence dated Sonora, October 13.
[5] Henry S. Brooks, "A Glance at California," *California Mountaineer*, I (1861), 14.

had driven out to inspect the tunneling operations at Table Mountain, Tuolumne County. While making the short trip out to the mountain, Brooks had found it necessary to stop his wagon repeatedly in order to alight and "open the well-made and substantial gates of the different farms." At the tunnel he had discovered that for living accommodations the canvas-roofed shanty of the middle fifties had been replaced by "a neat little house," in which the presence of a feminine cook was revealed by the "row of cups and saucers, and shining pans . . . suggestive of comfort and good cheer." [6]

Benjamin P. Avery, another observant mining-town editor, had a similar experience when he made a trip through the central section in September 1859. He traveled through districts in which, a few years before, he had seen only restless gold hunters living in flimsy hovels. Now he saw attractive houses, pleasantly set off by gardens and vines, and, best of all, he saw whole families where once there had been only men. [7]

The enthusiasm with which these contemporary descriptions were written should not mislead one into thinking that social normalcy had been entirely achieved. Conditions were improved in the sense that the majority of the mining towns were less boisterous and less exclusively masculine than formerly, and in the

---

[6] Henry S. Brooks, "The Romance of Table Mountain Tunnels. Rough and Ready," *California Mountaineer*, I (1861), 34-35.
[7] North San Juan *Hydraulic Press*, September 17, 1859.

sense that evidences of cultural life, such as debating societies and lyceums, were beginning to appear. There were, however, many local exceptions that tended to reduce the average rate of civilizing progress.

On the one hand there were areas such as that tributary to Columbia, where the attainment of a balanced society was less important than the indications of approaching decadence. It was the fate of towns located in such districts that caused the *Sonora Herald* to confess nervously:

> The future of towns built in the mining region of California is a problem that has elicited serious thought on the part of men doing business here, . . . It would be useless for us to deny that many persons entertain serious apprehensions as to the permanency of these interior towns, grounding their doubts on the probable exhaustion of the placers.[8]

On the other hand, there were towns in the northwest, such as Weaverville, at which society had advanced hardly at all from the conditions of 1850. When the state geological survey visited Weaverville in October 1862, its members counted twenty-eight saloons and found that "gambling and fighting are favorite pastimes." After witnessing three bloody street fights in succession, a French member of the survey wittily re-

---

[8] San Francisco *Daily Alta California*, December 20, 1858, clipping *Sonora Herald*. Cf. Henry S. Brooks, "The Future of our Mountain Towns," *California Mountaineer*, I (1861), 422-424.

marked: "I teenk dat de mineeng customs are petter preserved in dees plaze dan in any town I yet see in dees state." [9]

Then there were also a few boom towns, which relived in the late fifties and the sixties the days of 1849. Poverty Bar, for example, plunged into "regular '49 times, amid the general saturnalia of profligacy and its attending excitements," when some new gold deposits were uncovered in 1858. At Meadow Lake in 1866 conditions "forcibly reminded [one] of the flush times of '49 and '50." [10]

More important than any of these local exceptions was the general circumstance that the majority of the miners continued to be in the sixties what they had been in the fifties: unmarried men who lived in only moderate comfort and who displayed a tendency to nomadism and to a spendthrift existence. In 1861 John S. Hittell said of them:

Most of the miners live in a rough manner. . . . Not one-half of them lay up any money. Many earn money with ease, and spend it as fast as they make it. Men engaged in mining are not noted, as a class, for sobriety and economy. Their occupation seems to have an influence to make them spendthrifts, and

[9] William H. Brewer, *Up and Down California in 1860-1864. The Journal of William H. Brewer,* ed. by Francis P. Farquhar (New Haven, Connecticut, 1930), p. 330.

[10] Quotation on Poverty Bar from *New-York Daily Tribune,* December 27, 1858, clipping San Andreas *Independent;* on Meadow Lake from *Meadow Lake Morning Sun,* June 6, 1866.

fond of riotous living. Not more than one California miner in five has a wife and family with him. Most of the others are unmarried, and have no prospect of matrimony.[11]

A dozen years later, in 1873, the *Mining and Scientific Press* attempted to characterize the miners, in the hope of overcoming the unfavorable impression created by such descriptions as Hittell's. The miners were, the *Press* admitted, somewhat more wild and adventurous than other classes of men, and they were somewhat rougher because they were still comparatively isolated from the softening influence of feminine society.

Furthermore, those who lived at boarding houses or hotels in mining towns were apt to indulge in poker, seven-up, or cribbage as a means of escaping boredom in the evening. This did not, the *Press* cautioned, mean that the miners were desperate gamblers, nor did the previous admissions mean that they were constantly indulging in dissipation. On the contrary, most of them were sober, hard-working, hospitable men, even though "rough in exterior habit."[12]

In 1873 as in the fifties, the majority of the miners lived in their own rude cabins, on the outskirts of mining camps or in isolated localities near their claims. Such men still did their own cooking, with beans, potatoes, bacon, and bread as their staples. Their drinking and their celebrating they did on their occasional trips to the

[11] Hittell, *Mining in the Pacific States*, p. 211.
[12] *Mining and Scientific Press*, March 29, 1873, and May 17, 1873.

nearest town rather than at their cabins. The men employed by the large quartz companies were an exception, for they often lived at boarding houses or hotels.[13]

The appearance of the mining towns reflected the way of life of the miners. For the 1874 edition of his *Resources of California* Hittell used this description:

Most of them [i.e., the mining towns] are built with crooked streets through the middle of a cañon, which near the middle is densely lined with stores, billiard rooms, liquor shops, and restaurants. The dwellings are scattered about irregularly: some are neatly built and are surrounded with pleasant gardens; the majority are miserable little shanties or log-cabins, with no yard, flowers, or fruit-trees to give an appearance of home. The population is not permanent. One year the people are here, next they are elsewhere.[14]

Save for their lessened emphasis upon the riotous side of mining life, and save for their mention of the occasional presence of families and pretty homes, these descriptions could have been applied without drastic alteration to the men and towns of twenty years earlier, and yet the miners of 1873 were not just the miners of the early fifties grown older. Most of the men who came to El Dorado as participants in the great migration had, by the close of the fifties, disappeared from the chief mining centers. The death rate amongst them had been high. Some had returned home or had joined in the exodus into the newer mining frontiers. Others had turned to

[13] *Ibid.*, March 29, May 17, 1873.
[14] Hittell, *Resources of California*, p. 56.

319

new occupations in California, or to drink and despair. A few could still be found in remote parts of the mines or in decaying ghost towns.[15]

From the census statistics and from Hittell's estimate of 1873, it seems clear that the early miners were often succeeded by men of foreign birth. Of the latter, the Chinese were the most important. The census of 1860 showed that there were more Chinese in the state than there were immigrants from any other foreign nation. The census also showed that in the mineral regions the Chinese were ubiquitous. Each of the more important mining counties had two thousand or more of them.[16]

By 1873, unless Hittell was entirely wrong, the Chinese were the largest single racial or national group of miners, Americans included. This fact must be kept in mind when analyzing the character of the mining population of the later day, for it is quite apparent that the contemporary discussions instinctively excluded the Asiatics from their generalizations. What is even more regrettable is the almost universal failure of contemporary writers to leave behind them an adequate picture of the Chinese miners and their habits. The scraps of evidence that have survived show that the Chinese usually were relegated to inferior diggings, that they generally lived in inferior shanties, that they had few com-

[15] Cf. Forty-Nine (*pseud.*), "Where are the Forty-Niners?" *Hutchings' Illustrated California Magazine*, II (1857-58), 559-560; "Graves of the Forty-Niners," *ibid.*, III (1858-59), 133-134.
[16] *Eighth Census, 1860: Population*, p. 34.

forts, and that they subsisted on a limited diet in which rice, dried fish, and tea were the staples, and pork and chicken the luxuries.

They did not take kindly to American ways. Their dress, like their food, remained more Chinese than American. For a long time they retained the blue cotton blouse and the peculiar cotton trousers to which they had been accustomed in China. Their chief concession to western civilization was to purchase American-made boots that, for some reason, were always too large for them. Few learned to speak English with any fluency. Few became Christians. Few had much contact with the white miners.

In the cities and towns they generally congregated into their distinctive "Chinatowns," and in the mining districts they were very apt to live and work in bands. In either case many of their personal wants were supplied not through the normal American agencies, but rather through their own merchants and through the good offices of the several great "Companies." These last were powerful, highly organized mutual-benefit associations. Almost every Chinese belonged to one or the other of them, and relied upon his "Company" for hotel and hospital facilities when visiting San Francisco, for the loan of funds in time of need, and for postal service throughout the mines.

Like the miners of other races, the Chinese were sometimes guilty of an excessive amount of gambling, and of providing the brothels with a steady income. Unlike the other miners, they were rarely seen drunk in public. There were lurid tales of their opium smoking and other vices, but at this distance it is hard to distinguish where their detractors were speaking the truth and where falsehood. Some said they were honest; others that they were utterly unreliable. Some described them as being very cleanly about their persons; others as being disgustingly filthy.[17]

In most cases the Chinese were engaged only in placer mining with rockers, long toms, and river dams. The federal commissioner reported, however, that in three counties of the Southern Mines they were occasionally hired for quartz operations, and that "in many quartz-mines and stamp-mills throughout the West, Chinese labor is employed for certain inferior purposes, such as dumping cars, surface excavations, etc."[18]

At many of the quartz mines of the later era, the miners were chiefly Europeans. At Grass Valley, for example, most of the hired miners were Cornishmen. The citizens of that cosmopolitan little community included Germans, Irishmen, Jews, Frenchmen, Chinese, and

[17] The foregoing general description is based chiefly on: North San Juan *Hydraulic Press*, November 13, 27, December 11, 1858, February 5, March 12, 1859; Brewer, *Up and Down California*, p. 330; Hittell, *Resources of California*, pp. 40-47; Bowles, *Across the Continent*, pp. 238-247; Borthwick, *Three Years*, pp. 143-145, 262-267, 319; Horace Greeley, *An Overland Journey, from New York to San Francisco in the Summer of 1859* (New York, 1860), pp. 288-289.
[18] Raymond, *Statistics* (1872), p. 4.

Negroes, but "our mining is for the most part carried on by Cornishmen," the local newspaper declared.[19] At Sutter Creek, in the early seventies, the mining population was a mixture of Cornishmen, Irishmen, Austrians, and Italians.[20]

The foreign predominance that characterized the gold-quartz mines was repeated at the mines devoted to new or special minerals. In the quicksilver mines south of San Francisco Bay the laboring force was at first exclusively Mexican. Later Cornishmen, Chileans, and a few Irishmen were employed. At Mount Diablo many of the coal miners came from Wales, although some were from Pennsylvania. When Copperopolis was in its heyday, Cornishmen and Irishmen were numerous.[21]

In general, both in California and in other parts of the Far West there was a tendency for men of foreign birth to take the manual labor tasks and underground work, while Americans became specialists in handling machines and complicated equipment.[22] This trend was undoubtedly stimulated not only by the increasing mechanization of both hydraulic and quartz mining, but also by the

<hr>

[19] *Grass Valley National*, November 5, 1861. On Grass Valley's population see also *ibid.*, May 21, June 2, 7, 1864, January 3, February 13, 1865.

[20] Mason, *Amador County*, p. 112.

[21] William P. Blake, "Quicksilver Mines of Almaden, California," *American Journal of Science and Arts*, 2nd series, XVII (1854), 439-440; T. S. Hart, "Notes on the Almaden Mine, California," *ibid.*, 2nd series, XVI (1853), 139; Brewer, *Up and Down California*, pp. 139-143; Munro-Fraser, *Contra Costa*, p. 473; *Mining and Scientific Press*, October 26, 1863.

[22] King, "Introductory Remarks," *Tenth Census, 1880*, XIII, ix.

distaste which many Americans felt for employment as ordinary "hired hands."

The presence of so many foreign laborers aroused little comment, for this was an age which gave scant recognition to the working man's role in society. Occasionally a traveler remarked upon the quaint Cornish atmosphere of Grass Valley. According to one visitor, the conversation at breakfast in a restaurant might run like this:

"Hi say, coosen Jack, how is the work looken' now?"

"Oh, spara forthly, he be gone to peirch out all together, I thinks."

"Say, waiter, bring me two boiled heggs and a glass of 'arf and 'arf."

"Hi say, you waiter, do ee see this? Hit his ha bloody swindle; these heggs are not 'arf full." [23]

Aside from casual comments such as this traveler's, Californians paid little attention to the character of their mining population until the first real labor troubles in the history of the gold region burst suddenly into public prominence in the years from 1869 to 1873. There had been many so-called "strikes" in the latter half of the fifties, but they were in fact consumer boycotts rather than employees' threats to stop work. When the yield of the placers began to decline sharply after 1854, the many thousands of independent miners found that they

[23] *Mining and Scientific Press*, December 28, 1872.

could no longer afford to pay the water companies the high rates demanded for the use of ditch water. Their requests for a reduction in the rates were usually refused, because the water companies themselves were often in none too safe a financial position.

In order to force acceptance of their terms, the miners generally held public meetings of all those served by a given line of ditches. The action taken by the meeting was, in most cases, the adoption of a self-denying ordinance, under the provisions of which no one was to buy water from that particular company until the rates had been brought down to the desired level. A battle of will-power and financial strength then began between the company and the miners. After everyone had suffered severely from inactivity and hence from loss of income, matters were usually settled by a compromise which brought a partial reduction in the offending rates. Such "strikes" were voted in many districts in all three sections of the mines during the six or seven years that began in 1855.[24]

The general withdrawal from the shallow placers and from independent mining during the early sixties, removed the conditions which gave rise to these struggles between consumers and distributors. In the meantime,

[24] Sample strikes may be traced in: *Sacramento Daily Union*, March 17, November 24, 1855, March 21, 1856; *Sacramento Weekly Union*, April 5, 12, 19, May 24, 31, 1856, January 10, 24, 1857, April 30, November 26, 1859.

the foundations for a genuine mining labor movement were being laid on the other side of the Sierras, in the new silver region of Nevada.

An embryonic miners' union was established on the Comstock Lode in 1863 and was succeeded by a stronger organization in the following year. When the latter was driven out of existence by the bosses, its place was taken by a new "Miners' Union" which became so strong that it was able to enforce for many years the observance of a standard four dollar wage, despite a gradual reduction in the cost of living.[25]

Over on the California side of the mountains the success of the Nevada union was observed by the men who were working in the quartz mines, which were then the only employers of large groups of hired hands. At Grass Valley a controversy already existed between the working men and the superintendents and owners. Repeated tests had shown that by substituting the new dynamite, or "Giant Powder," for the black powder previously used, the cost of blasting could be materially reduced.[26]

The miners were bitterly opposed to the new explosive. They claimed that it created noxious fumes which caused headaches, nausea, and other discomfort. They were afraid that it would make possible a reduction in

[25] Lord, *Comstock Mining*, pp. 182-190, 266-268, 355-388.
[26] San Francisco *Daily Alta California*, July 9, 1869; *Mining and Scientific Press*, January 16, 1869.

the number of employees at each mine. For every charge of explosive a small hole had to be drilled in the rock. Because of the greater power of dynamite, the holes cut for it did not need to be as large as for black powder, and they could be drilled by one man working alone, instead of requiring the services of two as had been the custom with black powder. Colloquially this cheaper type of drilling was known as "single-handed," in contrast to the "double-handed" drilling used hitherto.

The dispute was the further complicated since in Cornwall, whence most of the Grass Valley miners had come, the quarrel over the relative merits of single-handed vs. double-handed drills was an ancient one that had sent its roots down deep into local customs and prejudices. Upon coming to America the Cornishmen had shed little of their inherited bias.

In protest against both dynamite and single-handed drills, the Grass Valley Cornishmen formed a miners' league and held meetings to consider action. The two leading mines, both managed by rugged veterans of that two-fisted industry, promptly retaliated by locking out three hundred and fifty men and advertising for one hundred single-handed drillers. There seems to have been little violence until the management attempted to reopen one of the closed mines with "scab" laborers, some of whom were not of Cornish extraction. In the resultant fighting several "scabs" were slugged into insensibility, and the alarmed citizenry organized a Law and Order Association to combat the miners' union.[27]

After a period of desultory warfare, the strike disappeared temporarily from public notice. The issues behind it remained unsettled and were destined to cause a more serious outburst in 1872. In the meantime, the union leaders managed to exercise at Grass Valley a three year rule which an unsympathetic local newspaper termed a "tyranny imposed by ignorance." [28]

For the moment the scene shifted to the other great quartz center, Sutter Creek. There a miners' league was organized under the leadership of a man who had had experience in the unions on the Comstock Lode. Most of its members were Irishmen, Cornishmen, Austrians, or Italians.

The first dispute between the league and the employers broke out in the summer of 1871. The immediate cause of it was wages. Now wages in the California mining industry had shown a decided stability after the middle fifties. In the late fifties the rate was about three dollars per day, with a frequent downward trend toward two dollars. When the United States commissioner circulated a questionnaire in 1870, he received reports which showed that there had been little

[27] The strike may be traced in: *Mining and Scientific Press*, April 17, May 22, 29, June 19, July 10, 17, 1869, February 24, March 30, 1872; San Francisco *Daily Alta California*, July 19, 1869; *Sacramento Daily Union*, May 28, June 14, 1869, March 18, 1872.

[28] San Francisco *Daily Alta California*, March 25, 1872, quoting an unnamed Grass Valley newspaper.

change. The replies to his inquiry generally stated that wages for first-class miners were three dollars per day, and for second-class two dollars and fifty cents. Surface laborers, who were sometimes Chinese, usually received two dollars.[29]

The precise form which the wage dispute took at Sutter Creek is not entirely clear. At first glance the league's published demands would appear to have involved nothing more than acceptance of the prevailing California standard of three dollars for "first hands," two dollars and a half for "second hands," and a ten hour working day. Apparently, however, the league meant to have surface laborers as well as second-class underground men included within the lower of the two rates, so as to establish a uniform minimum wage for all workers. This would have necessitated pay increases. Off to one side of the main dispute was the question of prohibiting Chinese labor. Be that as it may, the real issue soon came to be at Sutter Creek what it had tended to become at Grass Valley: not so much a question of terms of employment as of the implied right of the working men to a voice in the administration of the mines.

The psychology of the period and the industry was firmly set against tolerating any hint of employee rule. The president of the first mine to be approached by the league's representatives hurried down from his office in

[29] Raymond, *Statistics* (1870), pp. 13-83. In remote areas wages were sometimes $3.50 and $4.00. Chinese had formerly received $1.00 or $1.25, now $1.75 and $2.00. Raymond, *Statistics* (1872), p. 5.

San Francisco to give his personal veto to any concessions. The league then called a strike against all mines that would not meet their terms.

The general public in California was aghast at these developments. For the first time people began to scrutinize the composition of the laboring population, and suddenly they realized that they had in their midst a group of men who were chiefly of foreign birth and who were in every respect professional miners. They were men who had "little intercourse with the outside world" and who were governed by "their own codes of ethics and modes of thought," which had been inherited from generations of ancestors in the traditional European mining districts from which they had come.[30]

At the first sign of violence an appeal was made to the governor of the state for troops, on the ground that life and property were endangered. The governor responded by going to Sutter Creek to mediate. When his intervention failed, he ordered militia units into Amador County. For several weeks the soldiers did guard duty, while reporters for the metropolitan dailies kept the general public informed of exciting developments. This show of force enabled the mine owners to end the strike without making any substantial concessions save in regard to the use of Chinese laborers.[31]

[30] The quotation is from Mason, *Amador County*, p. 113; cf. *Scientific Press*, July 1, 1871.
[31] The strike can be traced in: *Scientific Press*, June 10, July 1, 15, 22, 29, August 5, 1871; *Sacramento Daily Union*, June 2, 7, 23, 24, 26, 28, 1871; Mason, *Amador County*, pp. 112-113.

Six months later, shooting warfare broke out at Grass Valley. It was preceded by unsuccessful negotiations between the league and the employers over the use of dynamite, and was accompanied at one mine by a demand from the Cornishmen for the removal of some Irishmen recently hired. This time the opposition was better organized than on the occasion of the strike three years before. Grimly the anti-league *Grass Valley Union* promised: "The citizens of the town are prepared to meet all emergencies. There will be no Amador war in this case. That sort of war requires a great deal of red tape. This war will be shorter and very much sharper than the Amador affair." [32]

The *Union* proved to be an accurate prophet. Some of the league sympathizers alienated public support by shooting at miners who tried to continue work, and talk of a citizens' vigilance committee immediately began. Just what action was taken the newspapers did not reveal, but they did make it clear that a month after the beginning of the strike the league went down in disastrous defeat. The league voted to withdraw its opposition to the use of dynamite, and a conservative San Francisco journal jubilantly reported that "the power of the organization has been broken, and many of those who were the chief encouragers of the former system of ruffianism have been compelled to leave." [33]

[32] *Scientific Press*, March 9, 1872, quoting *Grass Valley Union*.
[33] San Francisco *Daily Alta California*, March 25, 1872. For details of the strike see *ibid.*, February 6, 20, 26, March 14, 1872; *Scientific Press*, February 10, 24, March 2, 9, 23, 1872.

A year later there was one final strike at Sutter Creek. The cause of it was an attempt at the Lincoln mine, long owned by Leland Stanford, to force the miners to work longer hours on Saturday nights. Before it was settled this strike, too, produced fighting, shooting, and bodily injury.[34]

Nothing could have revealed more fully the change that had come over California mining than these ultra-modern strikes that marred its later years. The picture of drawn battle lines between strikers on the one hand and troops and corporate owners on the other might have been sketched out of the twentieth century. So vividly expressive was it of present-day urban industrial conditions that a stranger would hardly have believed that this was the same foothill region into which the brawling, independent, gold-hunting population had rushed a quarter of a century before.

Mining had started in California as a brave adventure, in which men of slight experience engaged with the help of a few partners. Now, twenty-five years later, it was an industry that was owned by investors in distant cities and was manned by professional miners of foreign origin. In 1848 it had offered to all the world the chance to be one's own master, and to all it had promised an equal opportunity for wealth. In 1873 its attractions were only those of an eastern factory town: employ-

[34] *Mining and Scientific Press*, February 1, 1873; *Sacramento Daily Union*, January 29, 1873.

ment at fixed wages and under set conditions, with no means of bettering them save by the hard road of trade unionism. One could almost say that within the short span of twenty-five years California mining had passed through a cycle that commenced with what the economists call "home crafts" and ended with what the socialists term "proletarian industry." Within that brief period California began, developed, and completed its industrial revolution.

333

XVIII

EPILOGUE

When the second number of the *Overland Monthly* appeared on the San Francisco newsstands in August 1868, it included among its offerings a story entitled "The Luck of Roaring Camp," by "F. B. Harte." Within the space of a half dozen double columned pages this little sketch told a sentimental yet vivid tale of the birth, brief life, and tragic death of a baby whose mother was a prostitute.

So frank a reference to a forbidden topic might perhaps have earned for the romance a momentary burst of attention in that era of literary circumlocutions, but it could hardly have given the story the nation-wide popularity that it so quickly won. What distinguished "The Luck of Roaring Camp" from other fictional writings of its time was the setting in which it was laid. The scene was a raw California mining camp of 1850. The actors were rough, bearded miners. The dialogue was picturesquely indecorous.

This was not Bret Harte's first venture into what critics now call "local color." As far back as 1860 he had tried his hand at tales that had a California mining camp for their setting. Not satisfied with the results, he abandoned the field for eight years.[1]

[1] George R. Stewart, Jr., *Bret Harte Argonaut and Exile, Being an*

334

His literary colleague in San Francisco, Mark Twain, followed much the same course. In 1865 Twain published a humorous anecdote called "The Notorious Jumping Frog of Calaveras County," based upon an incident that was supposed to have taken place at a camp in the Southern Mines of California. Despite the applause with which this was greeted, Twain did not for the moment attempt any further exploitation of this promising vein.[2]

It was not until the close of the sixties and the opening of the seventies that Harte, Mark Twain, the poet Joaquin Miller, and several lesser writers awoke to a recognition of the literary treasure that lay at their feet. Harte's "The Luck of Roaring Camp" was the first indication that there was a new western awareness of the possibilities of the western mining scene. "The Luck of Roaring Camp" was soon followed by a succession of similar mining-camp tales and poems by Harte, by Mark Twain's *Roughing It*, an amusing description of life on the Comstock Lode and in California, and by *Songs of the Sierras*, a slim volume in which Miller sang of

> The valor of these men of old—
> The mighty men of 'Forty-nine.[3]

Account of "The Luck of Roaring Camp," Condensed Novels, . . . (Boston and New York, 1931), pp. 105-107.

[2] Bernard DeVoto, *Mark Twain's America* (Boston, 1932), pp. 169-176; Franklin Walker, *San Francisco's Literary Frontier* (New York, 1939), pp. 193-195.

[3] Cf. Martin S. Peterson, *Joaquin Miller, Literary Frontiersman* (Stanford University, California, 1937), pp. 57-66.

The use of the colorful aspects of mining life for the purposes of fiction and poetry was thus postponed until after the mineral regions themselves had passed out of their period of picturesque boisterousness and into the later, more mature era when comparatively peaceful conditions prevailed. Apparently it was only after the flush days in the gold camps had changed from reality into folk history that they could take on the rosy glow of sheer romance. Apparently it was only when one could look back from a safe distance that the hardships and disappointments of life in the mines could be forgotten, that the all-important struggle to improve the techniques of the industry could be thrust aside in favor of the purely dramatic. Only then could one find readers for tales of miners who lived an attractively bizarre existence, and who drank, gambled, swore, and joked a great deal, but rarely mined.

It is not hard to understand why a writer, whose purpose is to entertain, sometimes selects and exaggerates in so unbalanced a fashion, but neither is it difficult to perceive that the narratives which result from such treatment are far from giving an accurate representation of the life they claim to portray. In so far as the changing fortunes of the miner are concerned, one could come nearer the truth by reading quite a different literary effort of the period, Henry George's *Progress and Poverty*, a book which has the somber theme of increasing want amidst increasing wealth. Or, for a picture of a special type of miner, one could read Prentice Mulford's reminiscent sketches, which tell much about the stubborn individualists who sought to continue their independent existence in the Southern Mines long after that way of life had ceased to be economically justifiable.[4] Books such as these contrast sharply with the fictional and near-fictional accounts, especially those by Bret Harte's imitators, but because the former are few, while the writings that stress "color" are many, a thick haze of romantic legend and mythology has settled over the California mining scene. This must be cleared away if one is to understand the significance of the quarter-century that began in 1848.

Within that span of twenty-five years, California mining passed successively through a short period of flush times, during which the rudiments of the trade were learned, then through a longer interval of transition, while the adjustment was being made to better methods of exploitation, and, finally, after the discovery of the Comstock Lode, through an era which was simultaneously a time of colonization beyond California's borders and of mature progress within the state itself.

By 1873 the trends that characterized this last epoch had been carried to a point of full development, and conditions had appeared which were in the sharpest contrast to those that had obtained in 1848, 1849, and 1850. In

---

[4] In addition to *Prentice Mulford's Story*, see Franklin Walker's collection of *Prentice Mulford's California Sketches* (San Francisco, 1935).

place of the pan, rocker, pick, and shovel, the miner now had sluices a thousand feet long, dynamite, the diamond drill, and a hydraulic instrument so powerful that no one knew its limits. The miner could now destroy mountains, where once he had struggled to dig away hillocks.

In place of the crude wooden stamp mill and the slow, mule-powered arrastre, he had now the efficient "California stamp mill" and its auxiliary, the steam-driven arrastre. Instead of wasting the greater part of his quartz-gold, the miner now used the chlorination process and the several mechanical devices that increased the proportion of gold saved. If he needed expert guidance for his operations, the miner of the later day could turn to the new state university, where the business of extracting minerals from the earth had been officially recognized as a science.

Where once the system of law and property ownership had been dependent solely upon the coöperation of one's neighbors, now it was founded upon the statutes of the United States and was enforceable in the courts of the state and federal governments. In the camps where Sunday had been a saturnalia and ladies as few as ministers, the miner could now find a reasonable degree of peacefulness and a goodly number of families, churches, and other evidences of normal life. If he were oppressed by his employers, the miner could, in the quartz industry at least, have some hope of the organized support of his fellow workers.

To be sure, there had been much loss along with the gains. Towns and districts had boomed noisily into prominence and then had disappeared as completely as dew before the morning sun. Whole counties had fallen into decline and were not likely to recover. And yet, if much of the mineral region was destined for oblivion, nevertheless its hectic life had not been useless. Vast states and territories in the West beyond the Great Plains had been pioneered by men who had learned their lessons in the old California camps, while within California itself the demands of the mining population had built up one of the nation's great cities—San Francisco—and had encouraged the development of an agriculture that by 1873 was making its influence felt in the wheat and wool markets of both Old World and New. Further still, the rise of population centers in California and the other mining commonwealths had been one of the several factors that inspired the railroad boom which in 1869 joined the Atlantic to the Pacific, and which soon afterwards opened the way to the full exploitation of the West by all forms of American enterprise.

Both in its direct and indirect results, then, the evolution of California mining was a phenomenon that had both intrinsic importance and significance in the broader field of American development. But when assigning to it a place of consequence in the nation's past, it is well to remember that this was an experience in which the whole world shared. The Gold Rush which inaugurated the mining era was in every sense an international movement that brought to California the ideas, methods, and men without which the gold deposits of the Sierras would have long remained little more than a local curiosity.

Nor did the contributions from outside the United States cease with the ending of the great migration. Cornishmen, Austrians, Italians, and Irish came in increasing numbers to work in the quartz mines, the Chinese labored wherever the lords of the land would permit, and in the quicksilver mines and the Southern Mines the Latin Americans were for many years an important element. Similarly, just as the stamp mill and arrastre of the early period were copied from Europe and Latin America, partly through the agency of the Georgia and Carolina gold regions, so in the later day the diamond drill was borrowed from France and the formula for dynamite from Sweden.

If, therefore, modern California and much of the West beyond the Great Plains owe their foundations to the mining boom which began in 1848, then they are decisively the product of the joint efforts of men from many lands. Thereby they become not less "American" but more so, for throughout her history the United States has been a country in which the course of development has not been confined to a narrow path cut by the descendants of a single racial stock, but rather has been routed over a broad thoroughfare dug by the hands of men and women of diverse origins. In California at least a dozen nationalities and half that number of racial strains made major contributions to the progress of mining, and the great state which flourishes today upon America's western border stands as a lasting monument to the effectiveness of their joint labors.

APPENDICES

APPENDIX A

GOLD PRODUCTION

Fiscal Year	Gold Produced in California
1848	$ 245,301
1849	10,151,360
1850	41,273,106
1851	75,938,232
1852	81,294,700
1853	67,613,487
1854	69,433,931
1855	55,485,395
1856	57,509,411
1857	43,628,172
1858	46,591,140
1859	45,846,599
1860	44,095,163
1861	41,884,995
1862	38,854,668
1863	23,501,736
1864	24,071,423
1865	17,930,858
1866	17,123,867
1867	18,265,452
1868	17,555,867
1869	18,229,044

345

Fiscal Year	Gold Produced in California
1870	$17,458,133
1871	17,477,885
1872	15,482,194
1873	15,019,210
1874	17,264,836

No contemporary records were kept of the amount of gold produced annually in California. The chief sources of information are the annual reports of the express companies as to the amount of bullion handled by them, the custom-house figures on bullion and coin exported from San Francisco, and the United States Mint statistics on deposits received. Most contemporary estimates were based on the express company and custom-house figures, especially the former.

J. Ross Browne, in *Reports upon the Mineral Resources of the U. S.* (1867), p. 50, made a table of the treasure shipments and added comments on the probable shortcomings of those figures. John S. Hittell, in *Mining in the Pacific States* (1861), p. 39, had previously made an attempt to estimate the annual production rather than the amount of gold shipped.

Louis A. Garnett, a San Francisco mining statistician, adopted a new approach, based upon the premise that the most reliable data were the mint statistics and the custom-house figures on exports of uncoined bullion. Since both categories of statistics were defective for the years prior to 1855, the element of estimate had to be large in the figures for the early years. For discussion of Garnett's hypothesis and contemporary opposition to it, see *Scientific Press*, March 18, 1871. Garnett's complete table, covering 1848-1883, was first published as an appendix to Bowie's *Practical Treatise* (table on page 288 of that work; explanatory letter, pp. 281-291). The figures given above are Garnett's.

Garnett's figures differ from those of Hittell chiefly in regard to 1848-49 and 1851-52. In the former period, Hittell makes the production $10,000,000 for 1848 and $40,000,000 for 1849. Considering the small size of the mining population prior to the latter part of 1849, and the crudity of early mining methods, this seems excessive. In the latter period, Garnett's figures exceed those of Hittell by more than $20,000,000 for each year (1851 and 1852). A part of the difference can probably be explained by Garnett's use of the fiscal year, for Hittell was apparently basing his estimates on the calendar year. If Garnett is correct, then the difference between the fat years of 1851-1852 and the leaner years that preceded and followed was distinctly sharper than well-informed contemporaries realized.

In 1896 Charles G. Yale, a leading mining authority and statistician, was asked by the state mining bureau to prepare a table of California gold production. After examining carefully all extant estimates and talking with their authors, he accepted Garnett's figures and republished them in California State Mining Bureau, *Thirteenth Report (Third Biennial) of the State Mineralogist, for the Two Years Ending September 15, 1896* (Sacramento, 1896), table inserted between pp. 64 and 65. Since then these figures have passed into standard use in so far as the state and federal mining bureaus are concerned. A convenient compilation, which reviews some of the weaknesses inherent in mining statistics, is that by James M. Hill, "Historical Summary of Gold, Silver, Copper, Lead, and Zinc Produced in California, 1848 to 1926," U. S. Bureau of Mines, *Economic Paper*, no. 3 (Washington, 1929).

Year	Daily Wage	Remarks
1853	$5.00	Beginning in 1852, miners were for several years wont to calculate their labor as worth $5 per day.
1856-1858	3.00+	Reports indicate that hired miners were receiving from $3 up to $4. Independent miners reported in 1856 as rarely able to exceed $3 without large capital outlay.
1859	3.00	Both hired and independent miners reported at $3.00 although one report said "mining laborers" at $2.50.
1860	3.00—	Some definite reports of hired mining labor at less than $3, and evidence that rate was nearer $2 in parts of Southern Mines, but rush to Comstock Lode raised wages somewhat before end of year.

The above table, intended to cover the long decline down to a relatively permanent wage level, has been compiled from reports appearing in contemporary newspapers, trade journals, books, and federal and foreign reports. The column of "remarks" indicates the difficulties encountered.

The wage level reached by 1860 was maintained without major change throughout the rest of the period. In the 1863 edition of his *Resources of California*, Hittell said that the "best" underground miners received

350

$4 a day and "common" miners $50 a month and board. The replies to the federal commissioner's questionnaire of 1870 generally stated that "first-class" miners were paid $3.00 per day and "second-class" $2.50. During the Sutter Creek strike of 1871 one of the mine owners stated that the mines had paid the same wages to employees for twelve years.

All of the above figures are distorted in one respect. Reports of wages in the quartz industry are both more frequent and more reliable than for placer mining, because of the greater use of hired labor and the greater stability of quartz. Quartz wages therefore bulk large in the foregoing table.

Within the quartz industry there was always a variation depending on whether the individual job was underground or on the surface, skilled or unskilled. For example, the *Sacramento Transcript*, May 29, 1851, quotes these rates for Nevada City: "drifters" $7.00 per day and board; windlass men, $5.00 and board; common laborers, $4.00 and board. There was also some variation between localities. Thus in 1870 miners in remote areas received as high as $3.50 and $4.00.

As noted in the table, in the early years the Mexicans tended to form a special element in the labor supply, paid at a lower rate than Americans and Europeans. Subsequently the Chinese were very definitely set apart in the wage scale. Raymond, *Statistics* (1872), p. 5, states that the Chinese were formerly paid only $1.00 to

$1.25 per day, but that they were by then able to demand $1.75 and $2.00, perhaps partly because of their greater experience with mining and partly because of greater experience with wage negotiations.

For purposes of comparing the cost of living with the miner's daily wage, it is interesting to note these two reports for 1852. At Chinese Camp, wages were said to be $5 per day, as against a charge of $8 or $9 per week if one ate at a boarding house, or $4 or $5 if he cooked his own meals. On the American River wages were quoted at $6, as against a weekly board charge of $10 (San Francisco *Daily Alta California*, May 15, 1852; *Sacramento Weekly Union*, August 28, 1852).

APPENDIX C

NOTE ON SOURCES

In its original form the annotated bibliography for this book fills one hundred typewritten pages. To reproduce so massive a document here would be an impossibility from a publisher's point of view, and a step of dubious value for all save a very few readers. Therefore no more will be attempted here than to list some of the more significant sources used. The footnotes in the text give a more detailed picture of the material consulted.

It might be well to suggest at the start that the best way to begin a study of California mining is to buy a good guidebook and make a trip through the foothills of the Sierra Nevada. Impressions gained in the mineral region can be strengthened by visits to the museums and collections maintained in San Francisco, Sacramento, and some of the larger mining towns.

Of printed records the following proved especially valuable:

I. NEWSPAPERS AND PERIODICALS:

For location see Winifred Gregory's *American Newspapers 1820-1936. A Union List of Files Available in the United States and Canada* (New York, 1937), and her *Union List of Serials in Libraries of the United States and Canada* (New York, 1927).

Grass Valley Telegraph, 1853-1858.

Grass Valley *Nevada National,* later *Grass Valley National* and *Daily National,* 1859-1865. The Grass Valley press was of especial importance because of the leading position of Grass Valley and Nevada County in mining, particularly in quartz.

North San Juan *Hydraulic Press,* 1858-1860. As a leading hydraulic mining town, North San Juan's weekly was important.

Sacramento Transcript, 1850-1851.

Sacramento Union, 1851-1873 (daily, weekly, and steamer editions). Sacramento was the supply center for the Northern Mines and the capital of the state. The *Union* was a well-edited, widely circulated newspaper known throughout the state and read "throughout the interior" counties.

San Francisco *Alta California,* 1849-1873 (daily, weekly, and steamer editions). In the early years the *Alta* was unrivaled and was especially noted for its reports from mining and farming counties. The fact that John S. Hittell was for many years on its staff gives it an especial importance.

Union (Arcata) *Northern Californian,* 1858-1860. Significant for the light it casts upon conditions in the northwest.

Individual issues, not constituting complete files, were consulted for other towns, especially Marysville, Stockton, Sonora, and Columbia.

California Mining Journal, 1856-1858 (weekly). Very important. The first mining journal. Largely filled with material prepared for the *Grass Valley Telegraph*.

Mining and Scientific Press, 1860-1873 (weekly). Title varies. Most important single source for mining history. Had wide and good coverage of the entire Far West but stressed California and Nevada.

California Mountaineer, 1861 (monthly).

Hesperian, 1858-1863 (semi-monthly, later monthly).

Hutchings' Illustrated California Magazine, 1856-1861 (monthly).

Overland Monthly, 1868-1873 (monthly). Of the foregoing, *Hutchings'* provided excellent material, the *Overland* was very good save that it began publication rather late for the purposes of this study, and the other two supplied occasional good articles.

2. FEDERAL AND STATE REPORTS:

Of contemporary reports the most important were the successive volumes issued by the Federal Mining Commissioner, beginning in 1867. These contain a wealth of statistical and factual data on Far Western mining.

Browne, J. Ross, and James W. Taylor, *Reports upon the Mineral Resources of the United States* (Washington, 1867).

Browne, J. Ross, *Report on the Mineral Resources of the States and Territories West of the Rocky Mountains* (Washington, 1868).

Raymond, Rossiter W., *Mineral Resources of the States and Territories West of the Rocky Mountains* (Washington, 1869).

———, *Statistics of Mines and Mining in the States and Territories West of the Rocky Mountains*. (Issued annually; Washington, 1870-1877).

Of the many good reports published since 1880 by the U. S. Geological Survey, the U. S. Bureau of Mines, and the California State Division of Mines (previously known as the State Mining Bureau), the following have been especially useful:

Johnston, William D., Jr., "The Gold Quartz Veins of Grass Valley, California," U. S. Geological Survey, *Professional Paper*, no. 194 (Washington, 1940).

Julihn, Carl E., and Frederick W. Horton, "Mineral Industries Survey of the United States: California. Calaveras County. Mother Lode District (South)," U. S. Bureau of Mines, *Bulletin*, no. 413 (Washington, 1938).

———, "Mineral Industries Survey of the United

States: California. Tuolumne and Mariposa Counties. Mother Lode District (South)," U. S. Bureau of Mines, *Bulletin*, no. 424 (Washington, 1940).

Knopf, Adolph, "The Mother Lode System of California," U. S. Geological Survey, *Professional Paper*, no. 157 (Washington, 1929).

Lindgren, Waldemar, "The Gold-Quartz Veins of Nevada City and Grass Valley Districts, California," U. S. Geological Survey, Seventeenth Annual Report, part II, in Secretary of the Interior, *Report*, IV (1896), 1-262.

———, "The Tertiary Gravels of the Sierra Nevada of California," U. S. Geological Survey, *Professional Paper*, no. 73 (Washington, 1911).

Logan, Clarence A., "Mother Lode Gold Belt of California," California State Division of Mines, *Bulletin*, no. 108 (November 1934).

Lord, Eliot, *Comstock Mining and Miners*, U. S. Geological Survey, *Monographs*, IV (Washington, 1883).

3. CONTEMPORARY BOOKS:

Bean, Edwin F., comp., *Bean's History and Directory of Nevada County, California. Containing A Complete History . . . with Sketches of the Various Towns and Mining Camps . . . Statistics of Mining* (Nevada City, California, 1867). Excellent material that is of more than local importance.

Blake, William P., *Notices of Mining Machinery and Various Mechanical Appliances in Use Chiefly in the*

Pacific States and Territories for Mining, Raising and Working Ores (New Haven, Connecticut, 1871).

Borthwick, J. D., *Three Years in California* (Edinburgh and London, 1857). One of the best of all the travel accounts.

Brewer, William H., *Up and Down California in 1860-1864. The Journal of William H. Brewer, Professor of Agriculture in the Sheffield Scientific School from 1864 to 1903*, edited by Francis P. Farquhar (New Haven, Connecticut, 1930). Brewer was Whitney's right-hand man in the Geological Survey.

Clappe, Louise A. K. Smith (Dame Shirley, *pseud.*), *The Letters of Dame Shirley. California in 1851-1852*, edited by Carl I. Wheat; Rare Americana Series, edited by Douglas S. Watson, nos. 5 and 6 (2 v.; San Francisco, 1933). Excellent on social conditions.

Cronise, Titus F., *The Natural Wealth of California. Comprising Early History; Geography, Topography . . . Mines and Mining Processes . . . Together with a Detailed Description of Each County* (San Francisco, 1868). An encyclopedia that is of great value for many topics.

Helper, Hinton Rowan, *The Land of Gold. Reality Versus Fiction* (Baltimore, 1855).

Hittell, John S., *Mining in the Pacific States of North America* (San Francisco, 1861).

———, *The Resources of California, Comprising Agriculture, Mining, Geography, Climate, Commerce*

. . . *and the Past and Future Development of the State* (San Francisco, 1863; many subsequent editions). Hittell was easily the most important contemporary writer for the purposes of this study. He was a careful and indefatigable collector of economic data, and the author of innumerable editorials, articles, and books. Usually his judgments were cautious and precisely stated. See index under Hittell.

Ingalls, Eleazer S., *Journal of a Trip to California, by the Overland Route Across the Plains in 1850-51* (Waukegan, Illinois, 1852).

Kelly, William, *An Excursion to California over the Prairie, Rocky Mountains, and Great Sierra Nevada. With a Stroll through the Diggings and Ranches of that Country* (2 v.; London, 1851). A first-rate account.

Marryat, Frank, *Mountains and Molehills or Recollections of a Burnt Journal* (London, 1855). A cheerful account of life in the mines.

Mason, Jesse D., *History of Amador County, California, with Illustrations and Biographical Sketches of its Prominent Men and Pioneers* (Oakland, 1881). Mason had witnessed many of the events he describes.

Moerenhout, Jacques A., *The Inside Story of the Gold Rush,* translated by Abraham P. Nasatir; California Historical Society, Special Publications, no. 8 (San Francisco, 1935). Letters of the French consul, 1848-49.

Mulford, Prentice, *Prentice Mulford's Story. Life by Land and Sea* (The White Cross Library; New York, 1889). Reminiscences of life in the Southern Mines.

Phillips, J. Arthur, *The Mining and Metallurgy of Gold and Silver* (London, 1867). Good descriptions by an experienced engineer.

Royce, Sarah, *A Frontier Lady. Recollections of the Gold Rush and Early California*, edited by Ralph H. Gabriel (New Haven, Connecticut, 1932).

Taylor, Bayard, *Eldorado, or, Adventures in the Path of Empire: Comprising a Voyage to California, via Panama; Life in San Francisco and Monterey; Pictures of the Gold Region* (2nd ed., 2 v.; New York, 1850).

Whitney, Josiah, *The Metallic Wealth of the United States, Described and Compared with that of Other Countries* (Philadelphia, 1854).

Wierzbicki, Felix P., *California As It Is & As It May Be. Or a Guide to the Gold Region*, edited by George D. Lyman; Rare Americana Series, edited by Douglas S. Watson, no. 8 (San Francisco, 1933). A sane description of California in 1849.

Woods, Daniel B., *Sixteen Months at the Gold Diggings* (New York, 1851). A faithful day-by-day journal, with some vivid scenes.

Yale, Gregory, *Legal Titles to Mining Claims and Water Rights, in California, under the Mining Law of Con-*

gress, of July, 1866 (San Francisco, 1867). In its field a classic.

4. SECONDARY BOOKS:

Bancroft, Hubert Howe, *Works* (39 v.; San Francisco, 1882-1890). The more one uses these fat volumes, the more he respects the achievement of Bancroft and his staff.

Bieber, Ralph P., *Southern Trails to California in 1849*, Southwest Historical Series, V (Glendale, California, 1937). Definitive.

Bowie, Augustus J., *A Practical Treatise on Hydraulic Mining in California. With Description of the Use and Construction of Ditches, Flumes, Wrought-Iron Pipes, and Dams; Flow of Water* (New York, 1885).

Caughey, John W., *California* (New York, 1940). A well written one volume survey.

Cleland, Robert G., *From Wilderness to Empire. A History of California, 1542-1900* (New York, 1944). An excellent picture of social and economic development.

——, and Osgood Hardy, *March of Industry*, in California, edited by John R. McCarthy (Los Angeles, 1929). An economic history of California.

Coy, Owen C., *The Humboldt Bay Region, 1850-1875. A Study in the American Colonization of California* (Los Angeles, 1929). A good study of a part of the remote northwest.

Howe, Octavius T., *Argonauts of '49. History and Adventures of the Emigrant Companies from Massachusetts, 1849-1850* (Cambridge, Massachusetts, 1923). An excellent analysis of the records of 124 companies.

Kemble, John H., *The Panama Route, 1848-1869*, University of California Publications in History, XXIX (Berkeley and Los Angeles, 1943). A thorough account of a major route to California.

King, Joseph L., *History of the San Francisco Stock and Exchange Board* (San Francisco, 1910).

Lindgren, Waldemar, *Mineral Deposits* (4th ed.; New York, 1933).

Lindley, Curtis H., *A Treatise on the American Law Relating to Mines and Mineral Lands within the Public Land States and Territories and Governing the Acquisition and Enjoyment of Mining Rights in Lands of the Public Domain* (2 v.; San Francisco, 1897). Ranks with Yale's book as a mining law classic.

Lyman, George D., *The Saga of the Comstock Lode. Boom Days in Virginia City* (New York, 1934). The best of the recent books on the Comstock.

McAdie, Alexander G., "The Rainfall of California," *University of California Publications in Geography*, I (1913-1917), 127-240 (Berkeley, California, 1914). Valuable statistics, graphs, and maps.

Morrell, William P., *The Gold Rushes*, in The Pioneer

Histories, edited by V. T. Harlow and J. A. Williamson (London, 1940). Mining "excitements" the world over.

Potter, David M., ed., *Trail to California. The Overland Journal of Vincent Geiger and Wakeman Bryarly*, Yale Historical Publications, Manuscripts and Edited Texts, XX (New Haven, Connecticut, 1945). The introduction is the best summary yet made of the various routes to California.

Rickard, Thomas A., *A History of American Mining*, A. I. M. E. Series (New York, 1932).

———, *Man and Metals. A History of Mining in Relation to the Development of Civilization* (2 v.; New York, 1932). Rickard writes from a background of years of experience in the mining profession.

Royce, Josiah, *California. From the Conquest in 1846 to the Second Vigilance Committee in San Francisco. A Study of American Character* (Boston and New York, 1886), in American Commonwealths, edited by Horace E. Scudder. A book that has great strengths and some decided weaknesses.

Shinn, Charles H., *Mining Camps, A Study in American Frontier Government* (New York, 1885). A "classic" study that is still of great value despite its excessive optimism and its dated hypothesis of the inherent capacity of the Germanic race for self-government.

———, *Land Laws of Mining Districts*, Johns Hopkins University Studies in Historical and Political Science,

2nd series, XII (Baltimore, 1884). Primarily a collection of mining codes.

———, *The Story of the Mine, as Illustrated by the Great Comstock Lode of Nevada* (New York, 1896), in The Story of the West, edited by Ripley Hitchcock. A good account, though not the equal of Eliot Lord's.

Trimble, William J., "The Mining Advance into the Inland Empire. A Comparative Study of the Beginnings of the Mining Industry in Idaho and Montana, Eastern Washington and Oregon, and the Southern Interior of British Columbia; and of Institutions and Laws Based upon that Industry," *Bulletin of the University of Wisconsin*, no. 638 (History Series, III), pp. 137-392 (Madison, Wisconsin, 1914). A very good study.

Walker, Franklin, *San Francisco's Literary Frontier* (New York, 1939). A fine study in local color literature.

Weatherbe, D'Arcy, *Dredging for Gold in California* (San Francisco, 1907). Some useful information on placer mining.

Wiel, Samuel C., *Water Rights in the Western States. The Law of Prior Appropriation of Water as Applied Alone in Some Jurisdictions, and as, in Others, Confined to the Public Domain, with the Common Law of Riparian Rights for Water upon Private Lands*

(3rd ed., 2 v.; San Francisco, 1911). A distinguished treatise.

Williams, Mary F., *History of the San Francisco Committee of Vigilance of 1851. A Study of Social Control on the California Frontier in the Days of the Gold Rush*, University of California Publications in History, XII (Berkeley, California, 1921). A painstaking study of the whole problem of extra-legal justice.

Wilson, Eugene B., *Hydraulic and Placer Mining* (New York, 1901).

Zollinger, James P., *Sutter: the Man and his Empire* (New York, 1939). An honest biography that cuts through the mist of legends and sympathetic accounts.

5. HISTORICAL REVIEW ARTICLES:

The following list is limited to seven articles that are cited in the text and that were of especial importance for this study.

Green, Fletcher M., "Georgia's Forgotten Industry: Gold Mining," *Georgia Historical Quarterly*, XIX (1935), 93-111, 210-228.

————, "Gold Mining: A Forgotten Industry of Ante-Bellum North Carolina," *North Carolina Historical Review*, XIV (1937), 1-19, 135-155.

Guinn, James M., "The Sonoran Migration," Historical Society of Southern California, *Annual Publications*, VIII (1909-1911), 31-36.

Raymer, Robert G., "Early Copper Mining in Arizona," *Pacific Historical Review*, IV (1935), 123-130.

————, "Early Mining in Utah," *Pacific Historical Review*, VIII (1939), 81-88.

Wright, Doris M., "The Making of Cosmopolitan California. An Analysis of Immigration, 1848-1870," *California Historical Society Quarterly*, XIX (1940), 323-343, XX (1941), 65-79.

Wyllys, Rufus K., "The French of California and Sonora," *Pacific Historical Review*, I (1932), 337-359.

Since 1947 many additional books and articles have appeared. The new guidebooks should perhaps be mentioned first. Under the direction of Olaf P. Jenkins, the State Division of Mines has issued an excellent volume that is a guide not only to the geology but also the surviving buildings: *Geologic Guidebook along Highway 49—Sierran Gold Belt. The Mother Lode Country* (California State Division of Mines, *Bulletin,* no. 141, San Francisco, 1948 and later printings). Muriel S. Wolle devotes a chapter to California in her illustrated guidebook, *The Bonanza Trail. Ghost Towns and Mining Camps of the West* (Bloomington, Indiana, 1953), pp. 106–147. *Sunset Magazine* has published *Gold Rush Country. Guide to California's Mother Lode and Northern Mines* (2nd ed., Menlo Park, California, 1963). The maps and photographs are up to *Sunset's* usual high standard, but the text is poor.

The present writer has reappraised the nature of California mining and its influence on the Far West in *Mining Frontiers of the Far West, 1848–1880* (New York, 1963).

Two admirable journals that give much insight into the daily life of the northwestern and southern mining regions, respectively, are: Doyce B. Nunis, Jr., ed., *The Golden Frontier. The Recollections of Herman Francis Reinhart, 1851–1869* (Austin, Texas, 1962), and Charles L. Camp, ed., *John Doble's Journal and Letters from the Mines, Mokelumne Hill, Jackson, Volcano and San Francisco, 1851–1865* (Denver, 1962). The present writer has edited a reprinting of a contemporary (1858) illustrated pamphlet: *The Miners' Own Book, Containing Correct Illustrations and Descriptions of the Various Modes of California Mining* (San Francisco, 1949). A special topic of major importance, although chronologically outside the immediate scope of the present volume, has been well discussed by Robert L. Kelley, *Gold vs. Grain: The Hydraulic Mining Controversy in California's Sacramento Valley. A Chapter in the Decline of the Concept of Laissez Faire* (Glendale, California, 1959).

On the Gold Rush, including the journey to California, an extraordinary amount of new material has been published since 1947, despite the very extensive coverage of that subject in earlier writings. In the centennial year Carl I. Wheat prepared a very helpful bibliography of significant books issued since 1848: *Books of the California Gold Rush. A Centennial Selection* (San Francisco, 1949). At approximately the same time John W. Caughey published a good general historical survey, *Gold is the Cornerstone* (Berkeley and Los Angeles, 1948), while in the present year (1964) we are promised a new study by J. S. Holliday, tentatively entitled *Pocketful of Rocks: A History of the California Gold Rush.*

APPENDICES

A major study of the overland routes to California was made by Dale L. Morgan while editing *The Overland Diary of James Pritchard from Kentucky to California in 1849* (Denver, 1959). This study should be compared to the editorial work of David M. Potter in connection with the Geiger-Bryarly journals, referred to in the preceding "Note on Sources." (And note that a new paperback edition has put the Potter volume back into print.)

Oscar Lewis reported on water routes in *Sea Routes to the Gold Fields: The Migration by Water to California, 1849–1852* (New York, 1949). Raymond A. Rydell discussed one of these routes in *Cape Horn to the Pacific; the Rise and Decline of an Ocean Highway* (Berkeley and Los Angeles, 1952).

The circumstances surrounding the original discovery at Sutter's mill were investigated by a team of archaeologists who reported their results in *California Gold Discovery: Centennial Papers on the Time, the Site and Artifacts* (California Historical Society, *Special Publication,* no. 21, San Francisco, 1947). Detailed research into the spread of the gold fever in California and throughout the United States was done by Ralph P. Bieber in "California Gold Mania," *Mississippi Valley Historical Review,* XXXV (June 1948), 3–28.

INDEX

Mills, D. O., Bank of, 167; president of Bank of California, 185

Miners, number of, 43, 243; description of, 69-70, 317-319; hardworking, 84; restless, 84-85, 176; daily life of, 86, 312-319; high cost of living for, 121-122; unemployed, 171-176; return to California, 286. *See also* Wages. *See also by nationalities:* Cornish; Chinese; Mexican; Austrian; etc.

Miners' juries, 203-204

Miners' meeting, adopts codes, 213-214

Mining, in medieval and modern times, 45-47; requires hard labor, 55-56; declining profitableness for average miner, 171-173; damage by flood and drought, 242; effect upon an area, 261-262; speculative character of, 265, 282-283; taught at University of California, 309

Mining, deep, defined, 147; first practiced, 147. *See also* Mining, tunnel; Mining, hydraulic; Tertiary gravels

Mining, drift, use of term, 150 *n*

Mining, hydraulic, need for cheap method, 150-151; origin of, 152-153; reduces cost, 154; delay in adopting, 155; requires enlarged water supply, 156, 161-162; canvas hose improved, 156-157; blasting cement, 157-158; use of stamp mills, 158; wastes gold, 159-160; under-current sluice, 160; tail sluicing, 160; causes unemployment, 175, 251-252; all-metal unit, 293-294; unfavorable to small capital, 297

Mining, placer, deposits formed, 41; early mining, 50; law lacks uniformity, 219-220; federal laws concerning, 232; gold production of, 286-287; compared with quartz, 286-287. *See also* Mining, river; Mining, hydraulic; Mining, tunnel

Mining, quartz, early, 130-131; description of, 132; mills described, 134-135; California stamp, 136-137; reduction in cost, 138; wastefulness, 139; use of arrastre, blankets, quicksilver, 140; metallic sulphides (sulphurets), 141; chlorination process, 142, 291-292; statistics, 143-145; revival, 144; compared with placer, 145, 286-287; failures, 145-146; ownership, 182; conventions to improve law, 217-219; federal laws, 231-232; California vein law vs. apex law, 236-238; in Amador County, 253-254; at Grass Valley, 258-260; limited to a few centers, 287; influenced by Comstock methods, 289-290; improved methods, 289-292; unfavorable to small capital, 297

Mining, river, first practiced, 60; defined, 124; speculative nature of, 124-126; on American River, 126; use of flumes, 128; peak of, 129; decline of, 130

Mining, tunnel, early losses in, 148; becomes important, 148; statistics of, 148-149; often unremunerative, 150; in later era, 286. *See also* Mining, deep

Mining, vein, *see* Mining, quartz

Mining camps, description of, 72, 79-80; location of, 79; Sunday in, 80-81

Mining codes, *see* Law, mining

Mining commissioner, *see* Commissioner of mining

Mining equipment, manufacture of, 186-187; export of, 187

Mining law, *see* Law, mining

Mining life, *see* Miners

Mining stocks, *see* Stocks, mining